Digital Transformation in Business and Society

"This is a great book full of information for those who want to better understand the digital revolution and how it is changing the whole of society, and business in particular. This book is a good guide to see this reality in a completely different way, through the scope of technology, and to better understand the exponential future we are living in."

—Dr. Besa Shahini, *Professor of Economics and Finance, University of Tirana, Albania*

"The book offers thoughtful, comprehensive insights into the far-reaching capacity of the digital footprint to reconfigure and remap every aspect of business and societal behavior. The authors have crafted a trailblazing compendium of digital transformation thought and action related to micro- and macro aspects of business and society. The book has the potential to become the defining work in this body of literature."

—Dr. Alvin Williams, *Professor and Chair, Department of Marketing, University of South Alabama, USA*

"One of the best books I have had the opportunity to read, I strongly recommend that you take it into account for your graduate programs."

—Dr. Sergio Garcia Agreda, *Director of the Master's Program in Digital Business, Universidad Privada Boliviana, Bolivia*

"This book is a timely text that will inform readers on a wide range of business and societal applications in the area. This text will help organizational psychologists understand the human factor in business decision-making and execution to then fully grasp the potential for digital transformation both in Business 4.0 and Society 5.0."

—Dr. Mervyn Jackson, *Head, Discipline of Psychology, RMIT University, Australia*

Babu George • Justin Paul
Editors

Digital Transformation in Business and Society

Theory and Cases

palgrave
macmillan

Editors
Babu George
Fort Hays State University
Hays, KS, USA

Justin Paul
University of Puerto Rico
San Juan, Puerto Rico

ISBN 978-3-030-08276-5 ISBN 978-3-030-08277-2 (eBook)
https://doi.org/10.1007/978-3-030-08277-2

This Palgrave Macmillan imprint is published by the registered company Springer Nature Switzerland AG.
The registered company address is: Gewerbestrasse 11, 6330 Cham, Switzerland

Contents

Notes on Contributors

Ogechi Adeola is a fellow/senior lecturer in the Marketing/Operations Department of Lagos Business School Pan-Atlantic University, Nigeria.

Frederick Avornyo is a doctoral student in the Department of Marketing and Entrepreneurship, University of Ghana Business School, Ghana.

Richard Boateng is an associate professor in the Department of Operations and Management Information Systems at the University of Ghana Business School, Ghana.

Meenu Chopra is an academician and researcher by profession and spirit. She is pursuing a doctorate program as a full-time research scholar from the prestigious Delhi Technological University (DTU) (formerly Delhi College of Engineering) under the DTU Fellowship program. She has worked as an assistant professor in the Department of Management Studies, Deen Dayal Upadhyaya College, University of Delhi, teaching BBS/BMS students, which is the only undergraduate management program run by the University. Her areas of interest include knowledge management, human resource management, and advertising and branding.

John Ekechukwu is a civil engineer leading transport infrastructure design at Turnbull Engineering, Sydney, NSW, Australia. He has recently organized and facilitated the Turnbull Engineering Transport Infrastructure Design School of Excellence in partnership with the University of New South Wales. Engineering leadership and development through digital transformation is one of his primary areas of interest.

Olaniyi Evans is a lecturer in the School of Management and Social Sciences of Pan-Atlantic University, Lagos.

Brian J. Galli works as Assistant Professor of Management Engineering at Long Island University—Post (LIU Post). He also owns Apex Strategies, Ltd, a company that specializes in continuous improvement consulting and training initiatives. The author's major field of study is continuous improvement and project management integration. He has over 11 years of experience in applying industrial engineering and continuous improvement tools and concepts in a wide variety of arenas, including health care, manufacturing, transactional, and service environments.

Babu George is an associate professor in the Robbins College of Business and Entrepreneurship, Fort Hays State University, KS, USA. His research examines various dimensions of technology as it interacts with business and society. Of these, human behavior in technology-mediated spaces constitutes one of his key focus areas.

Connor Greene is an undergraduate student in the Finance and Economics Department of Southern New Hampshire University, NH, USA.

Tony L. Henthorne is a professor and Associate Dean of Graduate Programs at the University of Nevada, Las Vegas. Technological innovations in the service industry and its marketing implications constitute one of his research streams.

Robert Ebo Hinson is a professor in the Department of Marketing and Entrepreneurship, University of Ghana Business School, Ghana.

İlayda İpek is a research assistant at the Faculty of Business, Dokuz Eylül University, Turkey. Her research interests primarily concentrate on international marketing, brand management, and consumer behavior. Her articles have appeared in the *International Business Review*, *Journal of Consumer Behavior*, and *Journal of Retailing and Consumer Services.*

Oluwamayowa Olalekan Iredele holds a PhD in Accounting. He is a postdoctoral research fellow in the School of Accounting, University of Johannesburg. His interest is in environmental management accounting and reporting.

Stanislav Ivanov is a professor and Vice Rector (Research) at Varna University of Management, Bulgaria. Ivanov is the Founder and Editor-in-Chief of the *European Journal of Tourism Research* and serves in the editorial boards of more than 30 other journals. His research interests include robonomics, robots in tourism/hospitality, revenue management, destination marketing, tourism and economic growth, political issues in tourism, and so on.

Sajal Kabiraj is a full professor at the International Business College, Dongbei University of Finance and Economics, Dalian, China. He has taught at postgraduate level in universities in Sweden, Austria, Malaysia, Canada, UAE, China and India. Information technology and digital inclusion is one of his primary areas of interest.

Maria Lai-Ling Lam is Professor of Business at Calvin College, Grand Rapids, MI, USA. She has been a senior consultant of AJ-Great Limited for several years. She has taught at two Christian universities in the United States for 14 years and several universities in Hong Kong and China for seven years She holds a PhD in marketing and organization behavior from the George Washington University, an MA degree in Religion Studies, and an MBA and a BBA from the Chinese University of Hong Kong. She has authored one book and 99 peer-reviewed articles and book chapters. Her areas of research interest are managing diversity and inclusion, yoga meditation, management education, leadership development, and corporate social responsibility.

Thorsten Lammers is a senior lecturer at the School of Information, Systems and Modelling, Faculty of Engineering and IT at the University of Technology Sydney, NSW, Australia. Informed by a previous career in strategy and technology management consulting, his main research interest lies in the impact of digital technologies on the future of work.

Yang Liu is an assistant professor in the Business Administration Department, Fitchburg State University, MA, USA. She holds a PhD in International Business, major in Marketing, from Southern New Hampshire University, NH, USA, and she also has a master's degree in Economics and Business from Sweden and a bachelor's degree in Accounting, major in Enterprise Financial Management, from China.

Mahmoud Abdulai Mahmoud is a senior lecturer in the Department of Marketing and Entrepreneurship, University of Ghana Business School, Ghana.

Tankiso Moloi is Professor of Accountancy at the School of Accounting, University of Johannesburg. He holds a PhD in Accounting and Finance as well as three master's degrees in Accounting, Financial Management, and International Relations. His interest is corporate governance, especially risk management.

Gabrielle Muniz is a freelance writer and editor. She recently graduated from Long Island University—Post (LIU Post), as she has lived in Long Island all her life. Gabrielle has extensive editing experience in the field of management engineering, with topics such as continuous improvement, project management, and change management. Aside from her freelance work, she has been writing many articles on topics that vary from dessert recipes to movie reviews for two online magazines: *Witty Serendipity* and *Her Track*. Overall, Gabrielle hopes to gain a fulfilling career in the writing industry.

Justin Paul is Professor of Marketing and International Business at Rollins College, FL, USA, and the Graduate School of Business, University of Puerto Rico. He is known as an author/co-author of text books such as *Business Environment* (4th ed) and *International Marketing, Export-Import Management* (2nd edition). Two books turned out to be best-sellers in Asia. He has served as a full-time faculty member with premier institutions such as the University of Washington, Nagoya University of Commerce & Business, Japan and Indian Institute of Management. Paul serves as a senior/guest/associate editor with the *International Business Review, Journal of Retailing & Consumer Services, Small Bus Economics, European Journal of International Management, Journal of Promotion Management,* and *International Journal of Emerging Markets.* In addition, he has taught full courses at Aarhus University, Denmark; Grenoble Eco le de Management, France; Universite De Versailles, France; ISM University, Lithuania; Warsaw School of Economics, Poland, and has conducted research paper development workshops in countries such as Austria, the United States, Croatia, and China. Paul has been a program director for training diplomats from different countries and served as a consultant to the Government of India. He has been an invited speaker at several institutions such as the University of Chicago, Fudan & UIBE-China, Barcelona, and Madrid, and has authored over 50 research papers and best-selling case studies with Ivey and Harvard. His articles have been downloaded over 300,000 times. He introduced the Conservative, Predictable and Pacemaker (CPP) model for internationalization and massage model/measure for brand management.

Anju Varghese Philip is a full-time doctoral student at the School of Management Studies, Cochin University of Science and Technology, India. Her research interests include human-computer interaction, specifically in the area of need satisfaction through technology use.

Sudipta Kiran Sarkar is an assistant professor at Sol International School, Woosong University, South Korea. The role of social media technologies in business is one of his major areas of research.

Stuti Saxena is a PhD student at the Central University of Haryana. Her research explores the use of information technology by governmental organizations.

Md. Nur Alam Siddik is an associate professor in the Department of Finance and Banking, Begum Rokeya University, Rangpur, Bangladesh. He holds a PhD degree in Finance from Dongbei University of Finance and Economics, Dalian, China. Digital finance and financial technologies is one of his primary areas of interest.

Asha Thomas is an assistant professor at Jagan Institute of Management Studies (JIMS), New Delhi, India. Her areas of research interest include knowledge management, organizational behavior, and marketing. She has about nine years of experience in teaching, as well as over three years of experience in the IT and telecom industry. She is pursuing a doctorate program as a part-time research scholar from the prestigious Delhi Technological University.

Michalis Toanoglou is the department chair and an assistant professor in International Department of Hospitality Management (SIHOM) of Sol International School (SIS) in Woosong University in South Korea. He has several years of experience as an executive in the hospitality industry in Greece.

Craig Webster is an associate professor in the Department of Management at Ball State University, Muncie, IN, USA. His research interests include the political economy of tourism, public opinion analysis, and event management. His most recent book is *Robots, Artificial Intelligence and Service Automation in Travel, Tourism, and Hospitality*, co-edited with Stanislav Ivanov. He teaches courses in hospitality management at Ball State University's Miller College of Business.

Kei-Wing Wong is a graduate student of Architecture, University of Michigan, Ann Arbor, MI, USA. She also has a Bachelor of Science degree from Calvin College, Grand Rapids, MI, USA. Her major was Environmental Science (Chemistry). Her minor was Architecture and Geography. She has worked for different sustainable cities projects. Her passion is to develop a universal and affordable design for people to flourish in hospitals and to cope with the threat of climate change.

Li Yang recently graduated from the master's program at W.H. Harrah College of Hotel Administration, University of Nevada, Las Vegas. His graduate thesis investigated how artificial intelligence might revolutionize the future of hospitality.

Zakkariya K. A. is Professor of Marketing and Human Resource Management at the School of Management Studies, Cochin University of Science and Technology, India, and is also the Director of DDU KAUSHAL Kendra, the Centre for Vocational Education in the university. With more than 20 years of teaching and research experience, he has been a member of various academic and administrative bodies of different universities in the country. He has more than 60 research publications in journals and books of national and international repute and is reviewer to many journals, including *Gender in Management* and *Journal of Management Development*.

List of Figures

List of Tables

CHAPTER 1

Introduction

Babu George

Introduction

Businesses are being transformed in fundamental ways as a result of the data revolution and societal drivers of change (Aral, Dellarocas, & Godes, 2013). The new era inaugurated by digital 2.0 technologies created a discontinuity and a fundamental shift in our ideas of business and society. Fundamental concepts that constituted the definitional parameters of business are undergoing a re-definition process. Technological revolutions brought to us increase in processing power, bandwidth, user-friendly interfaces with technology, the 'internet of things', to list a few benefits. Also, we live in a society that loves to consume radical change. These drivers together have transformed our situated existence in the human society (Uhl & Gollenia, 2016). For businesses, these changes mean the blurring of industry boundaries, modular business architectures, distributed innovation micro-structures, 'just-in-time' definitions of business performance, among others.

Most major businesses in our times are in the business of transforming themselves by the power of data and the technological assets that make it possible for them to use the data in the best possible ways (Matt, Hess, & Benlian, 2015). Systematic improvement in business performance is

B. George (✉)
Fort Hays State University, Hays, KS, USA
e-mail: bpgeorge@fhsu.edu

B. George, J. Paul (eds.), *Digital Transformation in Business and Society*, https://doi.org/10.1007/978-3-030-08277-2_1

achieved by means of re-engineering every aspect of the firm, including its external relationships, with technologies that promise scope and scale enhancements and also flexibility to create new business models (Davidson, 1993). There is a widespread realization that the path to competitive advantage is through data (McAfee, Brynjolfsson, Davenport, Patil, & Barton, 2012). In this backdrop, this volume brings together different perspectives from experts on how the data revolution has shaped different regions, industries, and domains. The authors in this volume have identified and distinguished the driving forces that make radical shifts in our society and businesses possible.

The Digital Transformation of Businesses and Societies

Knowledge has gained predominantly as a factor in contemporary consumption (Antons, Joshi, & Salge, 2018). Consumers are eager to be part of the knowledge sharing process—they co-create what they consume. In many situations, the boundaries between the buyers and the sellers are now permeant, observe Ranjan and Read (2016). Also, since most of the purchases are made online, it is vital to provide these consumers sensory experiences as close to what they would get while visiting brick and mortar shops. There is huge scope for innovation in this area and digital technologies fueled by the power of big data could provide the main lever for the same (Caputo, Evangelista, Perko, & Russo, 2017).

On the one hand, we have some evidence to support the claim that the digital revolution has helped us improve not just our business enterprises but our overall standard of living as a community of people, around the world. On the other hand, constant disruptions in the technology sphere need individuals and societies to pace up relentlessly. The digital divide is stark in certain regions of the world and also in certain demographic groups. Governments and the civil society have to come up with ways and means to bring the digital have-nots up to speed. This could be achieved by means of a combination of education, training, and other incentives. The private sector's role in this regard should be examined more, too.

Digital revolution has impacted and transformed almost every realm of human affairs (Makridakis, 2017). So, almost every current topic on business and society will have a certain aspect of it. When we look at issues with a thin focus, it is natural to see certain things as outliers. However, technology myopia encountered by their leaders has caused many businesses their very survival and we need to be mindful about unduly limiting the scope of the digital revolution (Vega-Jurado, Juliao-Esparragoza, Paternina-Arboleda, & Velez, 2015). It is impossible to make sense of our contemporary life without referencing it with technologies that laced the fabric of it (Brooke, 2016).

Will people work in the future? No, especially if we continue to define work the way we do it now (Cook, 2017). Are we ready for the new kind of work? What characteristics will define human workers and how will these be able to give them a competitive advantage over the robots? Will artificial intelligence evolve to the extent of making humans entirely replaceable in the workplace? These are but a sample from the cross section of the questions that futurists pose (Aronowitz, 1994; Collin & Young, 2000; Stiegler, 2018). These questions cannot be avoided when we discuss business implications of the fourth industrial revolution.

That said, for practical reasons, this volume needed to be delimited to certain key dimensions of it. The diagram given below is an attempt to distill the characteristic themes that the chapters contained in this volume represent. This is by no means an exhaustive or mutually exclusive list of themes. Several chapters cut across the theme boundaries—which, again, is a characteristic feature of the seamlessly networked lives of the digital natives that we are.

The key themes included in this volume are represented in Fig. 1.1.

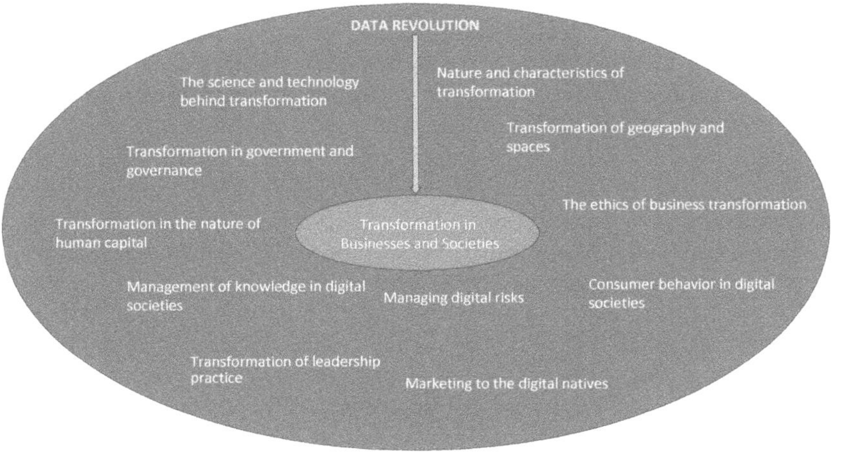

Fig. 1.1 Key themes included in this volume

A Brief Overview of the Chapters

Human ecological systems will not be the way we are currently familiar with them, as a result of our ability to manipulate big data (Graham & Shelton, 2013; Hampton et al., 2013). Ecology will not remain an esoteric pursuit, too. In the chapter titled 'Human Flourishing in Smart Cities', Maria Lai-Ling Lam calls for the importance of the development of enlightened citizens, rather than consumers, as the guardians of the smart cities managed by data-driven systems. As observed by Blagojević and Tufegdžić (2016), while technologies are disrupters, they also help preserve and refresh heritage of our geographical areas. 'The Making of Data Driven Sustainable Smart City Communities in Holiday Destinations' by Sudipta Kiran Sarkar, Michalis Toanoglu, and Babu George explains how data-centered technologies create smart and sustainable destinations for travelers. These chapters recognize the need for understanding digital technologies and data-driven systems as foundations of sustainable urban spaces.

Management of knowledge has gained a new meaning, as a result of our increased abilities to see insights in big data (Ding, Wang, & Wu, 2013). The industrial revolution 4.0 has brought in its associated phase of knowledge management 4.0 (Neumann, 2018). Asha Thomas and Meenu Chopra in their chapter 'On How Big Data Revolutionizes Knowledge Management' elaborate on this. To be able to see patterns as an emergent phenomenon needed newer heuristics and ways of thinking (Lv, Song, Basanta-Val, Steed, & Jo, 2017).

Social media is the frontline poster boy of digital transformation as the layman knows it. Social media strategy, more than social media technologies, drives the digital transformation directions of a vast majority of our organizations, observe Kane, Palmer, Phillips, Kiron, and Buckley (2015). While planning to revamp businesses through digital transformation, social media strategy cannot be let to the periphery (Westerman & Bonnet, 2015). Social media has become a central means by which to engage with the digital natives accustomed to this, and the decentralized nature of the social media makes engagement a complex enterprise (Stiver, Barroca, Minocha, Richards, & Roberts, 2015). The chapter 'Social Media Technologies and Export Marketing' by Frederick Avornyo et al. highlights aspects of it, in the particular context of changes in the age-old practice of exporting. Ogechi Adeola, Robert Ebo Hinson, and Olaniyi Evans, in their chapter 'Social Media in Marketing Communications: A

Synthesis of Successful Strategies for the Digital Generation', take a broader view and offer a synthesis of strategies for the digital generation.

Analytics is the backbone of what could be called a learning society (Larson & Chang, 2016). There is no exaggeration if someone calls it the religion of the digital society. In the chapter 'Data Analytics and Predictive Analytics: How Technology Fits in the Equation', Brian J. Galli and Gabrielle Muniz explain data analytics and predictive analytics through their various applications and examples of their benefits. Several insightful examples are provided to bring home the views. In a sector-focused chapter titled 'Artificial Intelligence and Robotics Technology in the Hospitality industry: Current Applications and Future Trends', Li Yang, Tony L. Henthorne, and Babu George explain the transformation of the hospitality industry, thanks in particular to advancements in robotics. All these chapters have a heavy emphasis upon the science and technology backbones of the data revolution that transformed businesses and societies.

Project management as an area of practice has changed dramatically, thanks to the power of information and communication technologies (Hoegl & Muethel, 2016). Strategy is traditionally distinguished from tactics and operations, in terms of time span and decisional level in the organizational hierarchy (Cascio & Montealegre, 2016). In the flat and fluid organizations in the networked economy, these distinctions have blurred away (Kahler, 2015). 'Digital Technology to Enhance Project Leadership Practice: The Case of Civil Construction' by John Ekechukwu and Thorsten Lammers portray how technologies increase the efficiency and effectiveness of project management, in the specific case of construction practices.

Organizations can now run on autopilot, except when major game changes are required (Goldspink & Kay, 2003). This also changes the role of people in organizations (Wirtz et al., 2018). To be a digital native is a radically new conception of life (Akçayır, Dündar, & Akçayır, 2016). In 'Effective Engagement of Digital Natives in the Ever-Transforming Digital World', Anju Varghese Philip and Zakkariya K. A. attempt to throw some light on how to manage digital natives both as internal as well as external stakeholders of the businesses. In the chapter titled 'Robotics, Artificial Intelligence, and the Evolving Nature of Work', Craig Webster and Stanislav Ivanov argue that humans will have to see their relationship to the job market differently and there will have to be an appropriate political response to the new economic landscape with changes in taxation and new ways of ensuring economic and political stability.

Big data does offer the promise of data-driven decision-making and change management. It has helped us identify subtle trends and relationships among key variables (Kshetri, 2009). Yet, the panacea has attracted so much unwanted attention from unscrupulous elements, too (Zhang, Xiao, Ghaboosi, Zhang, & Deng, 2012). Yang Liu, in the chapter titled 'The Dark Side of Big Data: Personal Privacy, Data Security, and Price Discrimination', shares some of these concerns. Dealing with cybercriminals needs new tactics and tools—many law enforcement departments are at the trailing edge of the curve in this regard. Given this, community policing has found a new reason for existence. Nonprofits and private sector corporations have taken the lead over government agencies in cyber defense—which, many consider to be problematic. On a related topic, 'Risk Management in the Digital Era: The Case of Nigerian Banks' by Tankiso Moloi and Oluwamayowa Olalekan Iredele examines an associated issue—the management of risks in the banking industry, using digital technologies.

Harford (2014) wonders if big data is a big mistake. Say, are we safe from computer hackers, and is the government doing enough to prepare? While the answer is not simple, this is a question worth asking, as also observed by Cuzzocrea (2014). Despite the risks, the inclusive process of knowledge creation and free flow of knowledge together could make our organizations significantly more responsive to stakeholder expectations and environmental cues (Boyd & Crawford, 2012). In 'Digital Finance for Financial inclusion and Inclusive Growth', Md. Nur Alam Siddik and Sajal Kabiraj elaborate on how technologies are levelers that could expand the inclusiveness of the financial sector.

Once upon a time, digital governance meant merely building websites for government agencies and using emails for communications (Pardo, 2000). West (2005) indicates that digital technologies have improved the efficiency and effectiveness of governments worldwide. Open data promises to open new vistas, expand our horizons, and stimulate innovation as never before. E-governance has replaced many traditional structures in governance, brought in transparency and interoperability (Gottschalk, 2009). It has made decision-making quicker and qualitatively better. 'An Evaluation of the National Open Government Data (OGD) Portal of the United Arab Emirates' by Stuti Saxena is an attempt to portray the success of one such initiative.

Current level of development of digital technologies offer new possibilities of thinking and doing. A particular issue that we need to overcome is the fact that, even as our societies have become data rich, they are still connection poor. 'Understanding Consumer Behavior in Technology Mediated Spaces' by İlayda İpek details the relations between consumers and sellers have transformed by means of various technologies that permit co-creation experiences.

Conclusion

Human beings were always social animals, but our wealth and its transfer could not have been embedded into the social relations until very recently. Digital currencies like bitcoins decentralized wealth from the clutches of the central banks and the nation states (Tapscott & Tapscott, 2016). Barter system is making a new homecoming, albeit with digital coins. The digital technologies have made innovations happen in unexpected corners of the world (Iansiti & Lakhani, 2017). Also, innovations have become more democratic and decentralized thanks to the 'networked brains' behind many of these innovations. The dissemination of innovations now happens at pace hither to unimaginable—actually, many new products launches now happen in the peripheral corners of the world, even before people in the developed countries get to buy them.

The increased choices online imply reduced loyalty (Connor, 2018). Cut throat competition also means reduced product lifecycles (De Bettignies & Duchêne, 2015). The existence of abundant online spaces for consumer activism means increased need for investing in customer relationship activities. Corporations increasingly employ big data analysis to gain key insights about consumers and markets. It is interesting to observe the challenges and solutions in digital marketing, as they evolve in increasingly digitized societies. In fact, most contemporary discussions on digital transformations directly or indirectly refer to digital marketing in general and social media marketing in particular. There is no better combined reflection of the impact of technological revolutions upon business and societies than social media and no wonder several chapters in this volume mentions social media in some way or other.

Digital technology integration into more traditional products means there is potential for disruption there, too (Yaqoob et al., 2016). Overall, digital technologies are enablers of disruption in every aspect of life. For instance, the idea of leadership is not anymore tied to hierarchy and structure. Knowledge leaders are a new breed. Isolated people sitting in different corners of the world but working together virtually as a team was unimaginable until recently. Geography is no longer a barrier. So also, is the traditional hierarchy built around mechanical principles: seamless cooperation and collaboration across organizations mean greater chances to synergize organizational resources.

With people spending significant portions of their lives online, criminals too migrated to the cyberspace (Saini, Rao, & Panda, 2012). Digital crimes do not end in financial fraud and malpractices. People's physical safety and security could seriously be compromised, too. Intrusions into the privacy of individuals have similar dimensions. Our increasing reliance upon big data results in the sidetracking of morals and meaning from the equations. Also, big data cannot effectively identify the micro-level nuances impacting grassroot levels—be it in organizations or in the civil society. The massive surge of data analytics as a profession is fueled by our need to tie everything to measurable data. If data science does not grow up enough to reveal the 'truth beyond data', many futurists believe we will be faced with stale robotic existence.

It is our collective amnesia that makes us think technological revolutions did not have a history. It is to be noted that mankind has used technologies ever since the dawn of civilizations. Most often, technologies have helped us advance—or, advances in technologies reflect our quest for leading better lives. However, there are also waypoints in our technological development wherein we made mistakes. Certain technologies not just failed, but also failed us. It is important to learn from these mistakes, as we design our future societies around a new set of technologies.

References

Akçayır, M., Dündar, H., & Akçayır, G. (2016). What makes you a digital native? Is it enough to be born after 1980? *Computers in Human Behavior, 60*, 435–440.

Antons, D., Joshi, A. M., & Salge, T. O. (2018). Content, contribution, and knowledge consumption: Uncovering hidden topic structure and rhetorical signals in scientific texts. *Journal of Management*. https://doi.org/10.1177/0149206318774619

Aral, S., Dellarocas, C., & Godes, D. (2013). Introduction to the special issue—Social media and business transformation: A framework for research. *Information Systems Research, 24*(1), 3–13.

Aronowitz, S. (1994). *The jobless future: Sci-tech and the dogma of work*. University of Minnesota Press.

Blagojević, M. R., & Tufegdžić, A. (2016). The new technology era requirements and sustainable approach to industrial heritage renewal. *Energy and Buildings, 115*, 148–153.

Boyd, D., & Crawford, K. (2012). Critical questions for big data: Provocations for a cultural, technological, and scholarly phenomenon. *Information, Communication & Society, 15*(5), 662–679.
Brooke, H. (2016). Inside the digital revolution. *Journal of International Affairs, 70*(1), 29–53.
Caputo, F., Evangelista, F., Perko, I., & Russo, G. (2017, September). The role of big data in value co-creation for the knowledge economy. In *10th Annual Conference of the EuroMed Academy of Business.*
Cascio, W. F., & Montealegre, R. (2016). How technology is changing work and organizations. *Annual Review of Organizational Psychology and Organizational Behavior, 3*, 349–375.
Collin, A., & Young, R. A. (2000). *The future of career.* Cambridge University Press.
Connor, J. M. (2018). Loyalty: The emotion of future expectation, felt now, based on the past. In *Emotions, everyday life and sociology* (pp. 37–51). Routledge.
Cook, N. (2017). *Enterprise 2.0: How social software will change the future of work.* Routledge.
Cuzzocrea, A. (2014, November). Privacy and security of big data: Current challenges and future research perspectives. In *Proceedings of the First International Workshop on Privacy and Security of Big Data* (pp. 45–47). ACM.
Davidson, W. H. (1993). Beyond re-engineering: The three phases of business transformation. *IBM Systems Journal, 32*(1), 485–499.
De Bettignies, J. E., & Duchêne, A. (2015). Product market competition and the financing of new ventures. *Management Science, 61*(8), 1849–1867.
Ding, G., Wang, L., & Wu, Q. (2013). Big data analytics in future internet of things. *arXiv preprint arXiv:1311.4112.*
Goldspink, C., & Kay, R. (2003). Organizations as self-organizing and sustaining systems: A complex and autopoietic systems perspective. *International Journal of General Systems, 32*(5), 459–474.
Gottschalk, P. (2009). Maturity levels for interoperability in digital government. *Government Information Quarterly, 26*(1), 75–81.
Graham, M., & Shelton, T. (2013). Geography and the future of big data, big data and the future of geography. *Dialogues in Human Geography, 3*(3), 255–261.
Hampton, S. E., Strasser, C. A., Tewksbury, J. J., Gram, W. K., Budden, A. E., Batcheller, A. L., … Porter, J. H. (2013). Big data and the future of ecology. *Frontiers in Ecology and the Environment, 11*(3), 156–162.
Harford, T. (2014). Big data: A big mistake? *Significance, 11*(5), 14–19.
Hoegl, M., & Muethel, M. (2016). Enabling shared leadership in virtual project teams: A practitioners' guide. *Project Management Journal, 47*(1), 7–12.
Iansiti, M., & Lakhani, K. R. (2017). The truth about blockchain. *Harvard Business Review, 95*(1), 118–127.
Kahler, M. (Ed.). (2015). *Networked politics: Agency, power, and governance.* Cornell University Press.

Kane, G. C., Palmer, D., Phillips, A. N., Kiron, D., & Buckley, N. (2015). Strategy, not technology, drives digital transformation. *MIT Sloan Management Review and Deloitte University Press, 14*, 1–25.

Kshetri, N. (2009). Positive externality, increasing returns, and the rise in cyber-crimes. *Communications of the ACM, 52*(12), 141–144.

Larson, D., & Chang, V. (2016). A review and future direction of agile, business intelligence, analytics and data science. *International Journal of Information Management, 36*(5), 700–710.

Lv, Z., Song, H., Basanta-Val, P., Steed, A., & Jo, M. (2017). Next-generation big data analytics: State of the art, challenges, and future research topics. *IEEE Transactions on Industrial Informatics, 13*(4), 1891–1899.

Makridakis, S. (2017). The forthcoming Artificial Intelligence (AI) revolution: Its impact on society and firms. *Futures, 90*, 46–60.

Matt, C., Hess, T., & Benlian, A. (2015). Digital transformation strategies. *Business & Information Systems Engineering, 57*(5), 339–343.

McAfee, A., Brynjolfsson, E., Davenport, T. H., Patil, D. J., & Barton, D. (2012). Big data: The management revolution. *Harvard Business Review, 90*(10), 60–68.

Neumann, G. (2018). Knowledge Management 4.0—Implications of the fourth industrial revolution on knowledge management in supply chains. *Theory and Applications in the Knowledge Economy*, Poland: Poznan University of Economics and Business 452–461.

Pardo, T. (2000). *Realizing the promise of digital government: It's more than building a web site*. Albany, NY: Center for Technology in Government.

Ranjan, K. R., & Read, S. (2016). Value co-creation: Concept and measurement. *Journal of the Academy of Marketing Science, 44*(3), 290–315.

Saini, H., Rao, Y. S., & Panda, T. C. (2012). Cyber-crimes and their impacts: A review. *International Journal of Engineering Research and Applications, 2*(2), 202–209.

Stiegler, B. (2018). *Automatic Society: The future of work*. John Wiley & Sons.

Stiver, A., Barroca, L., Minocha, S., Richards, M., & Roberts, D. (2015). Civic crowdfunding research: Challenges, opportunities, and future agenda. *New Media & Society, 17*(2), 249–271.

Tapscott, D., & Tapscott, A. (2016). *Blockchain revolution: How the technology behind bitcoin is changing money, business, and the world*. Penguin.

Uhl, A., & Gollenia, L. A. (Eds.). (2016). *Business transformation management methodology*. Routledge.

Vega-Jurado, J., Juliao-Esparragoza, D., Paternina-Arboleda, C. D., & Velez, M. C. (2015). Integrating technology, management and marketing innovation through Open Innovation models. *Journal of Technology Management & Innovation, 10*(4), 85–90.

West, D. M. (2005). *Digital government: Technology and public sector performance*. Princeton University Press.

Westerman, G., & Bonnet, D. (2015). Revamping your business through digital transformation. *MIT Sloan Management Review, 56*(3), 10.

Wirtz, J., Patterson, P. G., Kunz, W. H., Gruber, T., Lu, V. N., Paluch, S., & Martins, A. (2018). Brave new world: Service robots in the frontline. *Journal of Service Management, 29*(5), 907–931.

Yaqoob, I., Hashem, I. A. T., Gani, A., Mokhtar, S., Ahmed, E., Anuar, N. B., & Vasilakos, A. V. (2016). Big data: From beginning to future. *International Journal of Information Management, 36*(6), 1231–1247.

Zhang, Y., Xiao, Y., Ghaboosi, K., Zhang, J., & Deng, H. (2012). A survey of cyber crimes. *Security and Communication Networks, 5*(4), 422–437.

CHAPTER 2

Human Flourishing in Smart Cities

Maria Lai-Ling Lam and Kei-Wing Wong

Introduction

What cities *Homo sapiens* will choose to live will reflect the expectations of what our good life will be. The European Parliament (2014, p. 9) defined a smart city as "a city seeking to address public issues via ICT-based solutions on the basis of a multi-stakeholder, municipally based partnership." Smart cities, with all information and communication technologies (ICT), are assumed qualities to make us live well. Hollands (2015, p. 62) criticized that smart cities hid many issues, as information technology was assumed to "automatically make cities more economically prosperous and equal, more efficiently governed and less environmentally wasteful." Are we truthfully living happier in smart cities? Are we becoming a part of a predictable, efficient, and controllable system? Will we be disempowered and even lose our humanity in the highly rationalized system? What norms and values about a good life have been presented in the smart-city litera-

M. L.-L. Lam (✉)
Calvin University, Grand Rapids, MI, USA
e-mail: ml36@calvin.edu

K.-W. Wong
Michigan University, Ann Arbor, MI, USA
e-mail: wkei@umich.edu

B. George, J. Paul (eds.), *Digital Transformation in Business and Society*, https://doi.org/10.1007/978-3-030-08277-2_2

ture? How should we deconstruct these norms? Are there other alternatives for us to live happier in smart cities?

During the period between 2010 and 2017, we reviewed the literature, visited more than 40 international cities in Asia, Europe, and North America, interviewed their citizens about their respective cities, and explored the thinking of how we should lead our students in colleges to have critical engagement with the pervasive and affordable digital technology in their respective cities. We chose two leading smart cities in Asia to illustrate how economic prosperity and global competitiveness are used to frame their well-being. Singapore and Songdo in Republic of Korea were chosen, as their respective national governments were highly ranked in the area of provision of online services, telecommunication connectivity, and human capacity by the United Nations' 2016 E-government survey (United Nations, n.d.). Singapore and Republic of Korea both were named as Asian Dragons in the 1980s and aggressively adopted information technology to sustain its economic growth in the twenty-first century. They also compete with each other for global investments and are often ranked very closely in terms of smart cities development. Their governments have invested heavily in the development of smart city as their national industrial strategy (Central Policy Unit, 2015; Kim, Jung, & Choi, 2016; Team Finland Future Watch Report, 2017). They both inspire what future cities in China will be and what needs to be done for human flourishing in our higher education.

This chapter proceeds with five sections. Section "Socially Constructed Norms and Values of Smart Cities" describes the socially constructed norms and values of smart cities. Section "A Tale of Two Cities: Singapore and Songdo" describes Singapore and Songdo. Section "Critique of Smart Cities and Human Flourishing" includes the critiques of smart cities and human flourishing. Section "A Paradigm for Human Flourishing in Smart Cities" proposes a new paradigm of the development of smart cities that enhances human well-being. Section "Implications to Public Policy and Higher Education" discusses the necessary changes in our public policy and our higher education.

Socially Constructed Norms and Values of Smart Cities

The smart city has become popular when cities are aggressively marketed to create economic prosperity, global competitiveness, environmental sustainability by major ICT (information and communications technology) firms, engineering, property development, and construction companies (Hollands, 2015). The well-being of human lives will be improved in the densely connected, highly rational, and largely real-data analytic and programmable decision-making process in the smart cities (IBM, n.d.; Cicso, n.d.). White (2016) critically reviewed the marketing materials of smart cities provided by these global large ICT corporations, including Cisco and IBM, and learned that the technological intervention in the development of a smart city is legitimated by these three recurring crises: resource pressures brought on by rapid urbanization and an aging population, global climate change, and fiscal austerity and interurban competition for foreign direct investment and highly skilled workers. The need for the development of smart cities has been reframed as essential when the fear of quick urbanization, climate change, aging, meager public resources, and keen intercities competition are legitimated as the truth. Many issues related to data protection acts, inclusivity, costs, benefits re-distribution, and civil participation are not addressed in the rhetoric of technology-driven smart cities. The following section will explain how the development of smart cities has been legitimated and become part of our criteria for having a good life.

Characteristics of Smart Cities

The concept of smart cities is still quite vague. Smart cities are also named as intelligent, rational, green, sensor, wired, creative, sentient, digitized, or ubiquitous (u) cities...etc. The smart cities are equipped with all advanced technological devices such as the internet of things and multiple sensors and have become a promising discipline, industry, and academic discipline (Vanolo, 2014). Caragliu, Del Bo, and Nijkamp (2009, pp. 47–49) reviewed the literature about smart cities and summarized the characteristics of a smart city that is common to many research findings:

1. The utilization of networked infrastructure to improve economic and political efficiency and enable social, cultural, and urban development.
2. An underlying emphasis on business-led urban development.
3. A strong focus on the aim to achieve the social inclusion of various urban residents in public service.
4. A stress on the crucial role of high-tech and creative industries in long-run urban growth.

5. Profound attention to the role of social and relational capital in urban development.
6. Social and environmental sustainability as a major strategic component of smart cities.

Thus, a smart city relies on the use of the internet of things to improve economy and political efficiency. It mainly relies on business-led urban development through high-technology and creative industries. It provides many new business opportunities that lead to urban development. These opportunities are like Airbnb and Uber. It also emphasizes social inclusion, and social and environmental sustainability in the urban development. It provides technology-enabled participatory platforms such as New York City Big Apps. The smart-city proponents advocate the effective implementation and integration of smart technologies in a smart city will "improve quality of life, boost economic growth, mitigate the negative effects of climate change, and even foster more active citizenship" (Gaffney & Robertson, 2018, p. 49). People will flourish when there are synergies between economic growth and environmental protection in the smart cities.

Frost and Sullivan (2013), a global research and consulting company, estimated the size of smart-city market would be US 1.565 trillion by 2020. It defined a smart city consisting at least five of the following eight elements: "smart energy," "smart buildings," "smart mobility," "smart healthcare," "smart governance," "smart education," "smart security," and "smart citizens" (p. 2). Citizens are assumed to be better when they are smart enough to adopt various smart concepts or products in their daily lives. The smart-city industry can drive the development of other industries such as ICT, construction, communication, transportation, banking, insurance, health care, and education industry. Thus, smart city can be used as an instrument to improve national competiveness, which enhances community and quality of life. Kim et al. (2016) verified the positive impact of the smart-city industry on the national economy of Korea in terms of higher production, value-added, and employment induction effects. They learned that the key to the success of the smart-city industry was to acquire the competitive edge of the ICT and construction industries. The national wealth would be assumed to spill over into many people in different locations in the activated knowledge networks. They concluded that their study and numerous empirical studies have showed that "the various ICT applications in smart cities ultimately led to the build-up of intellectual capital or the creation of wealth by means of the interconnection of human and social capital via activated knowledge networks" (p. 16).

Chourabi et al. (2012) reviewed six working definitions of a smart city and concluded that a smart city is like a large organic system connecting many subsystems and components to collect, connect, and analyze real-live data in different urban domains for a better city. They proposed eight critical factors for the success of government-driven smart-city initiatives: management and organization, technology, governance, policy context, people and communities, economy, built infrastructure, and natural environment. Sadly, the current practices of smart cities in many countries are used to prioritize technology more than other factors.

Indeed, there is no clear shared definition or global consensus about the concept of smart cities. The concept of smart city can be easily manipulated as a government software-driven urban managerial tool and an opportunistic public policy (Cowley, Joss, & Dayot, 2017). There is no such research about people and communities in smart cities. Gaffney and Robertson (2018, p. 49) condemned the vague definitions and outcomes of smart cites as follows:

> This lack of clarity and consensus in the academic literature opens space for city government and ICT companies to apply the term 'smart city' flexibly according to their needs and desires. Furthermore, the stated aims of smart cities—such as sustainability, livability, intelligence—are often abstract goals that are difficult to measure and define in themselves. The abstract aims allow cities to declare smartness without great accountability.

Neirotti, De Marco, Cagliano, Mangano, and Scorrano (2014) analyzed 70 international cities' smart initiatives and learned that the main focus was in the area of transportation and mobility, natural resources, and energy. "Government is the domain in which the cities included in the sample report the lowest number of initiatives" (p. 18). Asian cities have exhibited a lower coverage in the smart initiatives related to human capital and government practices than European cities. They concluded, "cities that are more active in the domains that are aimed at improving their capacity to 'sense and act' through ICT systems are less likely to differentiate the initiatives launched for soft domains related to human capital, cultural heritage, and innovation" (p. 24). Without much investment in the human capital and government in the civil society, will the wealth creation in the industries in Asian smart cities be more concentrated than in European smart cities?

Logical Reasons for the Rise of Smart Cities

The rise of smart cities is framed as an effective strategy to solve many problems in the urban cities given the demand of new urbanism and the supply of inexpensive interconnected of things. The business-led technological innovations proposed by the smart city seemed to be a response to the failure in the city development in the 1980s in the United States: suburban sprawling, car-oriented cities, serving elites all at the expense of the public. The rise of the smart city is related to the concept of smart growth and new urbanism which originated in the United States in the 1980s and later transferred to Europe. Vanolo (2014, p. 887) stated, “New urbanism in planning was aimed at improving the urban environment and the quality of life in cities by promoting communitarian ideas and limiting urban sprawl, land consumption, and proliferation of forms of development inspired by the logic of the automobile and personal mobility.” Later, a group of researchers in Vienna ranked 70 medium-sized smart cities by six criteria: smart economy, smart mobility, smart governance, smart environment, smart living, and smart people. The ranking of cities in the European smart cities’ reports determine the competiveness of a country. There are many voices about smart cities such as World Bank Smart Cities Knowledge Silo Breaker (KSB) and European Union Smart Cities Assessment. The objective behind the development of smart cities in Europe is to compete effectively with rival global economics as China and India aggressively develop smart cities (European Parliament, 2014). Smart city is becoming a tool to sustain its own country’s economic growth and results in higher human capital and quality of life.

The supply of inexpensive processing devices, wireless coverage, rich analytic data, easy apps, and visualized output enable policy makers to estimate the risk level of each location and efficiently allocate resources. Many data in the cities are collected and translated to relevant information in a format of simple diagrams. The large information and communication companies are pursuing smart cities as their core strategies. For example, Siemens has “sustainable cities” logo, and IBM has “smarter planets and smarter cities” logo. They are assumed to provide many technocratic solutions to solve our pressing urbanization problems and even deal with natural disasters such as earthquakes. More and more professional conferences and initiatives are disseminated to use smart cities as our panacea. The rhetoric of smart cities is changing the criteria of measuring a city. A successful city needs to fulfill the criteria of being smart. Thus, smart cities may help many governments to achieve their social, environmental, and economic targets at lower cost. This is the testimony from the assessment of smart city in the European Union (European Parliament, 2014, p. 19):

> In view of the challenges associated with growing European urbanization, as well as the wider agenda to tackle economic recovery poverty, unemployment and environmental damage, *the Europe 2020 strategy incorporates a commitment to promote the development of Smart Cities throughout Europe and to invest in the necessary ICT infrastructure and human and social capital development.* Smart Cities may play a part in helping to meet the targets set out in Europe 2020 by adopting scalable solutions that take advantage of ICT technology to increase effectiveness, reduce costs and improve quality of life.

The Europe 2020 strategy believes that the incorporation of advanced ICT technologies in smart cities will improve global competitiveness, sustainability, diversification of functions (or authority), and quality of life because European cities will also use ICT to enhance human and social capital in the universities, industries, government, and civil societies. One Polish scholar told the author that the term "smart city" was mainly used by her government to access funding from the European Union and be aligned with the Europe 2020 strategy. Smart city can be easily politically manipulated for particular design and management of cities (Vanolo, 2014).

Why Do Smart Cities Matter?

By 2030, 60% of world's population will live in urban areas and 95% of urban expansion will take place in the developing world. Fresh water supplies, sewage, public health, and the living environment will be seriously affected by the rapid urbanization. China and India will build more and more smart cities to deal with increasing urbanization problems (Central Policy Unit, 2015). According to IHS Markit (2014) research, the number of smart cities will rise from 13 to 88 within 12 years. Some notable global large information and communication corporations aggressively pursue the concept of the "smart city." IBM has "smarter planet" and "smarter cities." CISCO has "the internet of everything for cities." These big corporations always promise to have an open, distributed, collaborative, and accessible global neural system. They are assuming leadership roles in the collaboration process among government, corporations, and civil society. They continue to play prominent roles into urban governance structure around the world. They advocate that the densely connected, highly rational, and largely real-data analytic and programmable decision-making process in the smart cities improve the well-being of their clientele.

The development of an inclusive, safe, resilient, and sustainable city is one of the 17 United Nation's sustainable development goals (United Nation, 2017). With the usage of technological innovations and the high density of cities, smart cities may be a good alternative to achieve the United Nation's sustainable development goals. The local government can collect sundry real-time data and monitor the urban system in a more centralized, holistic, and automated manner. Engineers, scientists, designers, and city planners can use the digital technology to model the realities and visualize better structures that use less energy and are more resilient to disaster. These abstract visual models can be manipulated and socially communicated to the public for the development of an inclusive, safe, resilient, and sustainable city.

The smart cities can provide the technology-enabled participatory platforms such as New York City Big Apps and Urban Mechanics initiated in Boston. These platforms are used by local governments to keep citizens informed, solicit input from citizens about improving the city, and collaborate with the citizens. Thus, smart cities can redefine how the local government connects with citizens, fosters thoughtful collaboration, and improves the internal function and quality of public service delivery. Desouza and Bhagwatwar (2014) examined the role of government and citizens in the 25 most populated US cities' technology-enabled participatory platforms and learned that there were four archetypes: citizen-centric and citizen data, citizen-centric and government data, government-centric and citizen data, and government-centric and citizen-developed solutions. They expect that there will be more citizen-initiated technology-enabled platforms that shape, implement, and evaluate public policy and the governance mechanism in smart cities.

A Tale of Two Cities: Singapore and Songdo

Over the last decade, there are favorable conditions for the investment in smart cities in Singapore and the Republic of Korea: "a centralized governance favoring shorter decision-making processes for public investments and more rapid development time for their executive, a higher economic development rate, a low political risk, and unique weather conditions that determine particular needs, with special regard to transportation systems" (Neirotti et al., 2014, p. 15). Singapore and Korea central governments are key drives for the development of smart cities as a global competitive industry (Central Policy Unit, 2015; Kim et al., 2016; Team Finland Future Watch Report, 2017). Will these two smart cities, Singapore and Songdo in Republic of Korea, allow people to flourish in all-efficient designs? Will these two cities treat citizens' involvement as necessary benefits or barriers? Are citizens engaged in the process of designing their own cities and making changes for better future?

Singapore

Singapore is a city state and has committed to develop its country and city around the internet. It has Singapore's iN2015 and wants to become an intelligent country (Arun & Yap, 2000). It is claimed to be the smartest city on the earth according to a 2016 UK research (Juniper Research, 2016) and has a high financial well-being but a low score in well-being according to the Gallup Study (2015). Singapore's smartness is mainly driven by its long-term development in an information technology policy (Central Policy Unit, 2015). During 1981–1985, the Singapore government funded information technology initiatives to computerize government ministries, improve public service, and produce a group of computer experts. During 1985–1990, the government developed a strong export-oriented information technology industry through its National Technology Plan. During 1991 to the present, the city/state has to be transformed as information technology permeates into every part of life. Hollands (2008) concludes that the basic objective of the development of smart Singapore is to enhance national competitiveness and the quality of life of citizens. Its success is through progressively institutionalizing information technology in the government, business sector, and civil society. When one author visited Singapore in 2011, one university professor told her, "We had to work harder and harder as our economy grew faster in the past 10 years." One smart-city engineer told the author in a smart-city conference in August, 2016 in Hong Kong, "In terms of smart city practice, European cities are doing much better than Singapore as there are more privacy protections and community involvements in Europe." One Singaporean also lamented, "We have many intelligent devices in the city. We have to pay a higher tax." Singaporeans experience much less chance to be consulted about smart-city strategy or develop citizen initiatives than Europeans do

(Council of the European Union, 2010). One expert who has developed smart cities in Europe and Singapore told the author that Singapore has just moved from its baby-stage smart-city development since it enacted the Personal Data Protection Act 2012. A group of Finland experts stated the main hurdle in Singapore's smart-city development was the lack of citizens' participation. They said, "Singapore's smart city endeavors are nearly entirely state-directed, and the lack of consultation with citizenry and other constituents could result in applications falling short of needs" (Team Finland Future Watch Report, 2017, p. 6).

Songdo, Republic of Korea

Songdo in the Republic of Korea is always claimed as an exemplar of smart cities. It is named as "the city of the future," "the world's smartest city," or "Korea's high-tech utopia" (Overview of Eco Cities, n.d.). This city is a symbol of synergy between economic growth and environmental protection. It is driven by the central government's ubiquitous city policy and system. It represents a new paradigm shift as it has been moved from the use of conventional infrastructure to an intelligent one (Shwayri, 2013). It is framed as a city that consists of the best of different cities in the world. There are combinations of commercial buildings, residential areas, parks, streams, animals, humans, and interconnected technologies. It is "smart. It is the model for next generation cities" (Songdo IBD website, n.d.). It was built from the reclaimed wetland from the Yellow Sea and was near the Incheon international airport starting in 1994 after numerous political fights between the Central government and the local Incheon City government during 1991 and 1994. During the financial crisis in 1997, the central government had set appropriate legislations to attract global companies and related foreign business using high-end technologies and to create a world-class city in Songdo. The local and municipal governments had to secure foreign investments and expand exports. The Incheon City had to choose to partner with the private sector for the development of buildings in Songdo. "In 2001, the local government had to turn to POSCO Engineering and Construction, a South Korea chaeobol, granting it the development rights with specific instructions to partner with an American developer to plan and build Sondgo International Business District (IBD), an area covering six square kilometers from the 53 square kilometers reclaimed from the Yellow Sea to build Songdo city" (Shwayri, 2013). New Songdo International City, Inc. was formed by New York

City-based Gale International (61%), POSCO E&C (30%), and Morgan Stanley Real Estate (9%). New Cities (2014) quoted the speech of the CEO of Gale International, Mr. Stanley Gale, from metropolis now: "We want to crack the code of urbanism, then replicate it. We want to build at least twenty Songdo ourselves: G20-Gale 20." Mr. Gale showed his ambitious dreams of producing 20 Songdo smart cities in the future. CISCO, a key information technology company, was chosen to connect and optimizes everything. The city was expected to be completed by 2013.

However, when the author and her research assistant visited the city and interviewed the marketing manager of New Songdo International City, Inc. in May, 2016, we were surprised there were many empty buildings and few residents in the park. The city had to be finished by 2020. The Gale International did not succeed to attract enough foreign investors in the district. The Korean government had to change the targets and to provide more infrastructure to attract residents to relocate in the city. People have to move in after the buildings are built. One developer of an information technology engineering company told us the company was blamed when it really installed smart devices because the city just wanted to look environmentally friendly to attract foreign investment. Many smart garbage collection bins were demolished because there were not enough maintenance and residents. The Songdo city is too big to fail. The government must use taxpayer money to finance this project and continue maintaining the area. During 2003 to 2013, the investment in the city was more than US$ 100 billion, while the budget amount was US$ 35 billion. The city is framed as a smart city with all computerized accessible buildings and electrical sensors such that the city can be responsive to the needs of residents. However, it was centralized developed and mainly constructed for the attraction of multinational investments (Townsend, 2013).

As more multinational investments are in China, Songdo city is now used to attract Chinese investors to reside in this city. The development of Songdo still needs daily huge investments, while South Korea would like to keep Songdo as an exemplar of a smart city. The cost of the removal of huge habitat land for birds and the resettlement of many fishermen near Yellow Sea have not been addressed in the Songdo's development report. The city was originally designed to enable foreigners or tiny proportion of local employees working for foreign corporations to enjoy a quality of life that was like the expatriates' Western cities. In summary, the key criteria to measure the success of this smart city is to attract foreign investments and increase its international competiveness. The quality of life is mainly measured in terms of economic success created by the development of smart cities. Kim et al. (2016) reviewed the negative critiques of Songdo and specified that Songdo is "criticized as being motivated by real estate initiatives, in which the 'smart' city label is reduced to the mere adoption of ICT features for the apartments" (para. 1 under section 2.3). Many civil voices about the establishment of the city were suppressed. The top-down approach only "promotes private business interests while ignoring social needs" (Shwayri, 2013, p. 43). This is the comment done by Sennett (2012, para 7):

> Songdo represents the stupefying smart city in its architectural aspect—massive, clean, efficient housing blocks rising up in the shadow of South Korea's western mountains, like an inflated 1960s British housing estate—but now heat, security, parking and deliveries are all controlled by a central Songdo 'brain'. The massive units of housing are not conceived as structures with any individuality in themselves, nor is the ensemble of these faceless buildings meant to create a sense of place.

Lessons Learnt from Singapore and Sondgo

Singapore and Sondgo both are driven by the centralized government policies for national competitiveness and economic interests. Sondgo in the Republic of Korea is established from scratch and is an assemblage of pre-existing urban imaginaries (Vanolo, 2014). The image of Songdo cannot relate to its history and culture. It is placeless. It can be reproduced and was implemented without paying attention to the social and political inequality in different context. The seamlessly connected world in Singapore and Songdo's virtual images ignores how the existing social and political structure limits many people's human capacities to make choices and to live well. It seems that the smart-city model can be reproduced and solve cities' problems without knowing the causes of those problems. The development of smart city continues a McDonaldization effect and just wants to have more people to be in the city. Citizens can work as sensors to help their government to identify potholes or other problems by using their smartphones. Citizens are treated as consumers and entrepreneurs in these smart cities. When cities are created to be responsible for being smart according to a defined criterion in relation to technological development, environmental protection, and economic attractiveness, the public policy makers can be easily tempted not to invest their resources in helping their citizens to increase their capacities to enjoy the benefits of smart cities. Sometimes citizens might be treated as commodities to fulfil the political agenda. Citizens have to become smart people in order to help their cities to have higher ranking and assessment of the smartness for a better global competitive score.

Critique of Smart Cities and Human Flourishing

Many smart cities are controlled by corporations, and there is no distinction between public and private space when data is collected everywhere (Boyer, 2015). The public space is coded and the privatization of public space is disturbing. Individual's right of privacy is violated when data protection act is not enacted. Cities are assumed to be efficiently managed through the centralized operating system, data analytics, big data, and urban informatics. The complexity of a city can be abstract and distanced as simple informatics. The urban development was skewed toward the positive side of technology and attracted global capital for making industrial cities. Many alternatives of promising urban development are ignored (Caragliu et al., 2009; Hollands, 2008). The management of urban development is saturated with technological devices instead of the development of social and human capital. The city of the future has been sold to the corporations without much public officers' accountability to their people. Once corporations finish their projects, they often leave these to ill-prepared local government to run and maintain with all advanced technologies.

Many citizens are vulnerable to cyberattack and ransoms without much government protection. It is difficult to attribute the cause of the problem when many sensors are interconnected and there are time gaps between the causes and results in environmental and social performance. No one has complete expertise about the function of the complex technological system adopted in the smart cities. There are multiple uncertainties, multiple normative frameworks, and multiple scenarios of future in the complex technological system (Sarlwitz, 2016). Citizens with less resources living in highly connected technologies are foreseen to be more vulnerable than before the implementation of smart cities.

Cowley et al. (2017) reviewed the role of the public in five prominent smart UK cities and found that "the public is positioned as part of the city's routine functionality, rather than assigned 'co-creational' agency" (p. 12). These five-star smart cities emphasized service-user publicness and entrepreneurial publicness and ignored political and civic publicness. The practice of smart city in the public policy was quite opportunistic and did not draw much attention from the public. The development of smart-city initiatives in the United Kingdom comes from the Department of Business, Skills, and Innovation, not from the Department of Communities and local government. Public policy makers are eager to use the smart cities' initiative to provide economic and political efficiency. Smart cities are "politically inclusive and culturally creative in only limited ways" (Hollands, 2008, p. 304).

There are serious problems of surveillance, discrimination, and freedom infringement in the development of smart cities. The big data collected by the smart cities can be a powerful tool to exclude or include particular data. The entire process of keeping or deleting particular data can be very political. There are significant power differences between knowledge of experts and citizen operators, the public and managers of information in these highly rationalized cities. Many data in smart cities are collected, coded, and operated with algorithms that are invisible or incomprehensible to the users and actors. The self-contained step-by-step set of operation in the algorithms is programmed without much transparency and public involvement. The algorithm may come from the set of data that affirms the existing social and political discrimination. It is questionable whether the public has the capability to meaningfully interact or engage in the discussion of the existing algorithm that governs the daily operation of the city. How can the existing algorithm be undergoing continuous improvement when so many people are incapable to criticize it even when the algorithm is claimed to be transparent to the public? It is the question who owns the data and who will not be tempted to sell the data. There are different social structures using different algorithms to make sense of data as information and form opinions about a person. There is information and power asymmetric in different social structures that can result is serious reputation losses to those who have less power. Sadly, those who cannot access the facilities in smart cities will be excluded from the network and may even bear a higher cost of living because the presence of many smart buildings and technology in their cities.

Rio de Janeiro (Brazil) can be a good case to illustrate how it was created in the idea of IBM and poorly managed by the local government. The city got the "2013 Smart City of the Year" in the Smart City Expo (Enbysk, 2013), is the exemplar of IBM's "smarter planet" and "smarter cities" marketing campaign. Gaffney and Robertson (2018) examined critically the implementation and functioning of two "smart cities" systems in Rio de Janeiro and learned that the smart technologies exacerbated pre-existing digital and social economic inequalities and served the rich more in limited wealthy areas. The notable security development Rio de Janeiro was around sporting venues, tourist areas, and key transportation infrastructures. The information collected was not used for long-term decision-making. More crimes happened outside the monitored areas. The smart technology was used to "consolidate the territorial, financial, and political gains realized during a major event" (p. 58). When there were protests in 2013–2014, the government used the smart technology to disrupt activist network. The smart technology only consolidates power, accumulates wealth, and generates honor for mayors and a few people through mega sports events in Rio de Janeiro! The smart technology does not solve the existing problems of radical inequality or systematic poor governance. Gaffney and Robertson (2018, p. 48) concluded as follows:

> the smart city paradigm is not capable of addressing the most pressing needs of cities with chronic deficits in urban infrastructure and an absence of robust civil society institutions. As such, smart city systems may actually contribute to the securitization and fragmentation of urban space, exacerbating social and political divides.

Another case is the authors' home town, Hong Kong. It was ranked fourth in the Asia Pacific after Seoul, Singapore, and Tokyo in the 2013 Cohen study of smart cities (Central Policy Unit, 2015) according to Boyd Cohen's definition of smart cities:

> Smart cities use information and communication technologies (ICT) to be a more intelligent and efficient use of resources, resulting in cost and energy savings, improved service delivery, quality of life, and reduced environmental footprint—all supporting innovation and low-carbon economy. (Footnote 3, p. 1)

Hong Kong is effective at using technology to improve operation efficiency in daily lives by adopting the Octopus electronic transaction system and its robust ICT infrastructure. The mobile and household broadband penetration rates are very high. However, the quality of life of people has dropped in the recent 10 years while the city is regarded as the world's second-most unequal city in terms of household income distribution (South China Morning Post, 2017, June 9). More and more people are complaining about the public policy and over-compact living environment. Although Hong Kong Planning department has advocated smart, green, and resilient strategy in the Hong Kong 2030+ document (Hong Kong 2030+ Website, 2016), there is not much public engagement and excitement about the policy. The public treats it as an opportunistic public policy. The 2030+ blueprint was criticized to run "counter to the goal of a true livable city" (South China Post, 2016, November 9). It was criticized not to develop actions to reach the standards listed in the 2030+ plan. Holloway (2017), CEO of Habit for Humanity Hong Kong, comments Smart Hong Kong Consultancy study only increases the income disparity as the inclusiveness of the city is missed in the 2030+ blueprint. The claim of smart-city development in Hong Kong cannot give comfort to the public who have to deal with serious economic and political inequality in their daily lives.

A Paradigm for Human Flourishing in Smart Cities

The new paradigm must be community-led urban development, not business-led urban development. The city is governed by the public who are empowered by technology. The development of smart cities should not be opportunistic in public policy and needs the involvement of the public and the awareness of the public about the values behind the decision that was made. This new paradigm embraces the investment in the soft domain of smart city including human capital and governance structure in Asian countries. Increasing the quality of people and inclusivity of the community should be the heart of the smarter city process. The government should develop data protection act to enact the rights of privacy of citizens.

ICT infrastructure and technology are used to empower the people and community to exercise their choices and to be engaged in the economic, social, and political activities in the smart cities. These technological companies need to inform citizens how their data are used and protected. Internet of people should exceed over internet of things. Multiple stakeholders must work within their perceived capabilities and values. They have to do civic experiments and also encounter technological problems in the limitations of our human brains and bodies. "But building smart cities is going to take time. It will by necessity be a long, messy, incremental process" (Townsend, 2013, p. 11). Developing smart cities for people needs to understand the perception of people about the cities and the priorities of values to precede the development. It values connectivity and practices empathy in a culture of care and justice. Human flourishing, human connectivity, and inclusiveness would be seriously addressed.

In this new paradigm, a sustainable flourishing city is derived from the social meaning of the people in the community (James, 2015). The vision of the future of the city should come from the people, not from those who sell the technology. The evolving expectations of households and neighbors about their cities will be captured in the community dialogues. In this new paradigm, the inhabitants' symbolic meaning and collective historical memory of the city are captured and highly advocated by Christopher Alexander in his numerous books (Alexander, 1979; Alexander et al., 1977; Alexander, Davis, Martinez, & Corner, 1985; Alexander, Neis, Anninou, & King, 1987). He advocates the new paradigm of human flourishing as the rediscovery of the lost "pattern language" from the hidden memories of ordinary people about the buildings and the cities. He explains this new paradigm in urban planning and building design as the timeless way and "quality without name":

> A building or a town will only be alive to the extent that it is governed by the timeless way....This quality in buildings and in towns cannot be made, but only generated, indirectly, by the ordinary actions of the people, just as a flower cannot be made, but only generated from the seed (Alexander, 1979, ix-xi).

Without the help of architects or planners, if you are working in the timeless way, a town will grow under your hands, as steady as the flowers in your garden (Alexander, 1979, 8).

For example, when Christopher Alexander designed Eishin School in Tokyo, he listened to the voices of all future users including principals, school teachers, and students; he also designed the building on the site based on the hand-on experience. Thus, the design of smart cities for people needs to be evolved slowly based on the needs of users and achieve "timeless way" and "quality without name" through the in-depth understanding of pattern languages commonly used in the city.

The new paradigm has been practiced by Lynch (1960) in his book titled *The Image of City*. He has visited the places and interviewed the residents of Boston, New Jersey, and Los Angeles cities. He has found that people in urban cities orient themselves through the means of mental maps, which provide them emotional security. The designer needs to understand how the environment is constructed through the interrelations of different groups and individuals and provides obvious structure with open patterns that allows the incorporation of participants' more complex mental representations of cities:

> Cities are the habitat of many groups, and only with a differentiated understanding of groups and individual images and their interrelations can an environment be constructed that will be satisfying to all until such knowledge is at hand, the designer must continue to rely upon the common denomination, or public image, and otherwise provide as great a variety of types of image-building material as he can devise. (p. 157)

> A city must have both an obvious structure that can be grasped immediately and also a potential structure which allow one gradually to construct a more complex and comprehensive picture. (p. 158)

In his book titled *What Time Is This Place* (Lynch, 1972), he describes how our innate sense of time affects the way we view and change our physical environment, especially in cities. He said, "Since we cannot be certain what will be most relevant in the future, we have an obligation to save some characteristic evidence of every major period—to establish an environment archive." To practice this new paradigm is to enable people to be connected to the historical meaning of the place and also to allow for future new developments.

The new paradigm is human-centered development and accepts our community-led urban development within our social and ecological limits. Ordinary people need to be equipped with the skills to know how to use the technology and to be included. They also need to be educated to know the pattern, quality, symbolic images, and mental representations of their cities when they are empowered to govern their own cities. Thus, the community-led growth is related to the grass-root smart growth movement that promotes sustainability by design and inclusive engagement from the public. "Sustainability emphasizes finiteness, limits, and ways of life that will advance that priority, rather than a continued emphasis on pursuing traditional forms of growth, but in a 'smarter' manner" (Beatly & Collins, 2000, p. 297). Thus, the symbolic meanings of the city need to be translated in a comprehensive human-centered city vision. The vision has to be articulated to the citizens, end-users, finance and corporate ICT firms, engineering, property development and construction companies, urban planners, and government.

The vision of the city is evolving and the design of the city for people comes from doing. Technology is only one of the alternatives that affects the future prosperity and development of human communities. Each individual is allowed to build the city which promotes environmental friendly habits by creating walkable neighborhoods containing a wide range of housing and job types. The government needs to be regulated by the community and then develop expertise to know how to equip the communities to develop their own cities for being resilient (Jacobs, 1961). Newman, Beatley, and Boyer (2009) urge cities to be prepared for short-term contingencies such as alternative transport availability, alternative fuel programs, and household awareness programs. Without these preparations, cities will begin to crumble and fall apart.

In summation, in the new paradigm of the development of smart city for people, we need to be patient with the complexity and ethical uncertainties in the process of developing a sustainable flourishing community that is aligned with existing social capital and structure. Policy makers must develop governance structure to protect citizens' data privacy. They also need to mobilize more engagement from the public, develop a process for social reflections about the values behind the public choices, need to develop contingent consensus, and learn by doing in the policy making. The new paradigm also challenges the major corporate ICT firms, engineering, property development and construction companies, and government to develop the city for people, not for quick financial interests and economic competitiveness.

Implications to Public Policy and Higher Education

Through the research of smart cities and the Asian exemplars, including Singapore and Songdo, we affirm the importance of development of citizens rather than consumers in our teaching and research. We need to facilitate our college students to practice community-led urban growth and embrace their civil responsibilities. We need to help our college students to wisely navigate through a vast ocean of digitized urban technology and solutions (Eshelman, Lam, & Cook, 2012; Lam, 2017; Vaill, 2007). We have to invest more smart initiatives in people and community, and governance in our home town, Hong Kong. It is essential for Hong Kong policy makers to develop trusting relationships with their citizens through responsible data policy, open and collaborative governance structure, and heavy investment in human resources if the economic benefits of a "smart city" are expected to spillover and improve the quality of life of local citizens. Citizens flourish when they can live in cities with dignity and wisely exercise their human capacities, including rational, emotional, bodily, sensory, relational, imaginative, and language-using abilities. Their basic capabilities of humanity need to be nurtured and respected before they can exercise their freedom and make their own choices in smart cities.

Here are some suggestions to Hong Kong policy makers.

Responsible Data Policy

Need to address the issues of data ownership, data protection, data interpretation, data regulation, and data security in the Data Policy. Citizens have rights to be deeply informed about how their personal data is collected and is selected for particular decisions. Data providers should know their risks of losing their reputation when their personal data have been chosen to form opinions about themselves. The open data, big data, the chosen algorithms used in the smart-city development should be managed with integrity and be accountable to citizens. The local government needs to learn from the best practices of other smarter cities and practice a citizen-focused data policy (Eurocities, 2016). The local government needs to strengthen its internal government and provide e-government solutions for the citizens. For example, citizens can access the real data about the gas emission of our monopolized utility companies.

Open and Collaborative Governance Structure

Need to foster community-level social capital that allows citizens to develop their own cities and treasure the symbolic meaning of the city with the support of technology. Cities should not be written in the language of a few powerful ICT companies and real-estate developers. They need to practice transparent governance and allow citizens to be partners, collaborators, and co-creators in the design and implementation of the city that has her "quality without name," and comprehensive human-centered city vision. Leaders from the private and public sectors need to examine what can be measured, monitored, and addressed in the smart city having significant moral implications to humanity. For example, the tender specifications of private projects in the development of smart cities need to be consulted. The public and private parties have to be accountable for the quality of lives of citizens who have to live with the consequences of the projects.

Heavy Investment in People and Community

They need to develop robust civil society institutions; need to reach out to, engage with, and empower people to find better solutions for urban problems and develop communities with compassion; need to educate people to be creative users of those ICT solutions for better community and quality of life; need to invest people's basic knowledge and skills of using those technologies; and need to allow people to question what "smart city" models are being used and what social norms and codes that guide the practices for the smart cities. For example, Hong Kong 2030+ plan should allow people to have numerous feedback about the proposed smart-city system and question the public values behind the plan. People should be empowered to develop social capacities for reflection about the social norms or algorithms for the interpretation of data as information, and participate in the political processes balancing the needs of the community, local government, and corporations in the development of smart cities.

In the development of smart cities, citizens require new public services, regulations, and accountability from policy makers and corporations. A city in which people reside is the smartest city when people care and work on it. "We want cities that work well enough, but are open to the shifts, uncertainties, and mess which are real life" (Sennett, 2012). Hirschman (1965, p. 35) reminded us of our high degree of initial ignorance and uncertainty when we implement the technological innovations to solve the problems of the city. He said,

> The term "implementation" understates the complexity of the task of carrying out projects that are affected by a high degree of initial ignorance and uncertainty. Here "project implementation" may often mean in fact a long voyage of discovery in the most varied domains, from technology to politics.

Educators need to enrich students' interpretations of these visual data of urban city solutions and guide them to ask many questions about human well-beings in smart cities. These are some suggested questions: (1) What does it mean for us humans to live well in smart cities? (2) What various issues can arise from an account that makes room for such a variety of human lives in smart cities? (3) What practices are used that undermine the flourishing of our families, our communities, our political groupings, and our culture in smart cities? (4) Will we continue to use these open-ended human capacities to construct valuable lives that yet are not seen in smart cities?

References

Alexander, C. (1979). *The timeless way of building.* New York: Oxford University Press.

Alexander, C., Davis, H., Martinez, J., & Corner, D. (1985). *The production of houses.* New York: Oxford University Press.

Alexander, C., Ishikawa, S., Silverstein, M., Jacobson, M., Fiksdahl-King, I., & Angel, S. (1977). *A pattern language: Towns, buildings, construction.* New York: Oxford University Press.

Alexander, C., Neis, H., Anninou, A., & King, I. (1987). *A new theory of urban design.* New York: Oxford University Press.

Arun, M., & Yap, M. (2000). Singapore: The development of an intelligent island and social dividends of information technology. *Urban Studies, 37*(10), 1749–1756. Retrieved from http://www.jstor.org/stable/43196456

Beatly, T., & Collins, R. (2000). Smart growth and beyond: Transitioning to a sustainable society. *Virginia Environmental Law Journal, 19*(3), 287–322.

Boyer, C. (2015). *Public space and the smart cities in the 21st century.* Speech delivered in Hong Kong Polytechnic University on November 27, 2015.

Caragliu, A, Del Bo, C., & Nijkamp, P. (2009). Smart cities in Europe. In *3rd Central European Conference in Regional Science—CERS* (pp. 45–59).

Central Policy Unit. (2015). *Research report on smart city.* The Government of the Hong Kong Special Administrative Region. Retrieved from June 20, 2017 from http://www.cpu.gov.hk/doc/en/research_reports/CPU%20research%20report%20-%20Smart%20City(en).pdf

Chourabi, H., Nam, T., Walker, S., Gil-Garcia, J. R., Mellouli, S., Nahon, K., … Scholl, H. J. (2012). Understanding smart cities: An integrative framework. In *2012 45th Hawaii International Conference on System Science (HICSS)* (pp. 2289–2297). IEEE.

Cisco Website. (n.d.). Smart+Connected Communities: Embrace digital innovation to create new revenue and better serve your citizens. Retrieved June 19, 2017, from http://www.cisco.com/c/en/us/solutions/industries/smart-connected-communities.html

Council of the European Union. (2010). Europe 2020—Public consultation: Overview of responses. Retrieved June 20, 2017, from http://ec.europa.eu/eu2020/pdf/overview_responses.pdf

Cowley, R., Joss, S., & Dayot, Y. (2017). The smart city and its publics: Insights from across six UK cities. *Journal of Urban Research & Practice*, 1–25. https://doi.org/10.1080/17535069.2017.1293150

Desouza, K. C., & Bhagwatwar, A. (2014). Technology-enabled participatory platforms for civic engagement: The case of U.S. cities. *Journal of Urban Technology, 21*(4), 25–50.

Enbysk, L. (2013, November 27). Rio de Janeiro wins top smart city honors. *Smart Cities Council.* Retrieved June 20, 2017, from http://smartcitiescouncil.com/article/rio-de-janeiro-wins-top-smart-city-honors

Eshelman, G., Lam, M., & Cook, M. (2012). Three contributing factors to effective utilization of technology in management education and practice: Personhood, Mindfulness, and Meditation. *Journal of the North American Management Society, 6*, 24–34.

Eurocities Website. (2016). Smarter cities: City-led, citizen-focused. Retrieved June 21, 2017, from http://nws.eurocities.eu/MediaShell/media/EUROCITIES%20stmt_smarter%20cities_June2016.pdf

European Parliament. (2014). *Mapping smart cities in the EU study.* Retrieved November 7, 2016, from http://www.europarl.europa.eu/RegData/etudes/etudes/join/2014/507480/IPOL-ITRE_ET(2014)507480_EN.pdf

Frost and Sullivan. (2013). Strategic opportunity analysis of the global smart city market: Smart city market is likely to be worth a cumulative $1.565 trillion by 2020. Retrieved June 21, from http://www.egr.msu.edu/~aesc310-web/resources/SmartCities/Smart%20City%20Market%20Report%202.pdf

Gaffney, C., & Robertson, C. (2018). Smarter than smart: Rio de Janeiro's flawed emergency as a smart city. *Journal of Urban Technology, 25* (3), 47-64. Retrieved July 24, 2019, fromhttps://doi.org/10.1080/10630732.2015.1102423

Gallup Study. (2015). Singapore ranks high in financial well-being, lags in other measure. Retrieved November 7, 2016, from http://www.channelnewsasia.com/news/singapore/singapore-ranks-high-in/1941492.html

Hirschman, A. (1965). *Journeys toward progress: Studies of economic policy-making in Latin America.* New York: Anchor Books.

Hollands, R. (2008). Will the real smart city please stand up? Intelligence, progressive or entrepreneurial? *City, 12*(3), 303–320.

Hollands, R. (2015). Critical interventions into the corporate smart city. *Cambridge Journal of Regions, Economy and Society, 8*, 61–77.

Holloway, H. (2017, November 1). Homes and hope: Smart cities. *Ming Pao.* Retrieved November 7, from https://news.mingpao.com/pns/dailynews/web_tc/article/20171101/s00017/1509473399797

Hong Kong 2030+ Website. (2016). Smart, green, and resilient city strategy. *Planning Department.* Retrieved June 19, 2017 from http://www.hk2030plus.hk/document/Hong%20Kong%202030+%20A%20SGR%20City%20Strategy_Eng.pdf

IBM Website. (n.d.). Smarter Cities: New cognitive approaches to long standing challenges. Retrieved June 19, 2017, from https://www.ibm.com/smarterplanet/us/en/smarter_cities/overview/

IHS Markit. (2014). *Smart cities to rise fourfold in number from 2013 to 2025.* Retrieved October 26, 2016, from http://press.ihs.com/press-release/design-supply-chain-media/smart-cities-rise-fourfold-number-2013-2025

Jacobs, J. (1961). *The death and life of great American cities.* New York: Random House.

James, P. (2015). *Urban sustainability in theory and practice: Circles of sustainability.* London and New York: Routledge.

Juniper Research. (2016, May 22). Juniper names Singapore as "smartest" city globally. Retrieved November 7, 2016, from https://www.metering.com/news/juniper-names-singapore-as-smartest-city-globally/

Kim, K., Jung, J., & Choi, J. (2016). Impact of the smart city industry on the Korean national economy: Input-output analysis. *Sustainability, 8*, 649. Retrieved June 19, 2017, from http://www.mdpi.com/2071-1050/8/7/649

Lam, L. (2017). Empathetic leadership as an alternative paradigm for responsible supply chain management. In N. Meunjohn & A. McMurray (Eds.), *The Palgrave handbook of leadership in transforming Asia.* London: Palgrave Macmillan.

Lynch, K. (1960). *The image of the city.* Cambridge, MA: MIT Press.

Lynch, K. (1972). *What time is this place.* Cambridge, MA: MIT Press.

Neirotti, P., De Marco, A., Cagliano, A. C., Mangano, G., & Scorrano, F. (2014). Current trends in smart city initiatives: Some stylised facts. *Cities, 38*, 25–36.

New Cities. (2014, December 28). South Korea conceptualizes the ultimate smart city. Retrieved November 7, 2017, from https://newcities.org/cityquest-songdo-south-korea-conceptualized-ultimate-smart-sustainable-city/

Newman, P., Beatley, T., & Boyer, H. (2009). *Resilient cities: Responding to peak oil and climate change.* Washington, Covelo and London: Island Press.

Overview of Eco-cities Website. (n.d.). Focusing future. Retrieved June 19, 2017, from http://focusingfuture.com/reader/Overview-of-Eco-Cities.html?page_n27=5

Sarlwitz, D. (2016). Innovation, the Techno-human condition, and ethics in radical uncertainty. Presentation at the *Sixth International Society of Business, Economics, and Ethics (ISBEE) World Congress*, July 13–16, Shanghai, China.

Sennett, R. (2012, December 4). No one likes a city that's too smart. *The Guardian.* Retrieved June 17, 2017 from https://www.theguardian.com/commentisfree/2012/dec/04/smart-city-rio-songdo-masdar

Shwayri, S. (2013). A model Korean ubiquitous eco-city? The politics of making Songdo. *Journal of Urban Technology, 20*(1), 39–55. Retrieved July 24, 2019, from https://www.tandfonline.com/doi/full/10.1080/10630732.2012.735409?scroll=top&needAccess=tre.

Songdo IBD Website. (n.d.). Smart. Retrieved June 21, 2017, from http://songdoibd.com/about/#smart

South China Morning Post. (2016, November 9). A green Hong Kong in 2030? Not according to this plan. Retrieved June 19, 2017, from http://www.scmp.com/comment/insight-opinion/article/2044045/green-hong-kong-2030-and-beyond-not-according-plan

South China Morning Post. (2017, June 9). What hope for the poorest? Hong Kong wealth gap with record high. Retrieved June 19, 2017, from http://www.scmp.com/news/hong-kong/economy/article/2097715/what-hope-poorest-hong-kong-wealth-gap-hits-record-high

Team Finland Future Watch Report. (2017). *Singapore, the 1st smart nation by 2025*. Retrieved June 20, 2017, from https://www.tekes.fi/globalassets/julkaisut/Future_Watch_Singapore.pdf

Townsend, A. (2013). *Smart cities: Big data, civic hackers and the quest for a new utopia*. New York: W.W. Norton & Company Ltd.

United Nations. (2017). Goal 11: Make cities inclusive, safe, resilient and sustainable? Retrieved February 3, 2017, from http://www.un.org/sustainabledevelopment/cities/

United Nations Website. (n.d.). UN E-Government Survey 2016. Retrieved June 19, 2017, from https://publicadministration.un.org/egovkb/en-us/Reports/UN-E-Government-Survey-2016

Vaill, P. B. (2007). Organizational epistemology: Interpersonal relations in organizations and the emergence of wisdom. In E. Kessler & J. Bailey (Eds.), *The handbook of managerial and organizational wisdom* (pp. 327–355). Thousand Oaks, CA: Sage Publications.

Vanolo, A. (2014). Smartmentality: The smart city as disciplinary strategy. *Urban Studies, 51*(5), 883–898. Retrieved November 7, 2017, from http://citeseerx.ist.psu.edu/viewdoc/download?doi=10.1.1.850.4942&rep=rep1&type=pdf

White, J. M. (2016). Anticipatory logics of the smart city's global imaginary. *Urban Geography, 38*(4), 572–589.

CHAPTER 3

On How Big Data Revolutionizes Knowledge Management

Asha Thomas and Meenu Chopra

Introduction

The present century being a knowledge economy, there is significant reliance on information and the availability of intellectual capital (IC). In this scenario, knowledge management (KM) has emerged to become an integral part of every organization. Increasingly, organizations are relying on the knowledge possessed by their employees for decision-making processes. Erstwhile KM practices and platforms were driven toward capturing and codifying the available explicit knowledge into digital artifacts, which made the collective organizational knowledge available to a much wider set of practitioners. These KM practices and tools were extremely successful across diverse industries, particularly in knowledge-driven sectors such as Consulting, Information Technology (IT), life sciences, and others. However, they were limited in that they were primarily focused on codifying the explicit knowledge artifacts and providing services for searching and retrieving these artifacts from the database.

A. Thomas
Jagan Institute of Management Studies, New Delhi, India

M. Chopra (✉)
Delhi Technological University, New Delhi, India

B. George, J. Paul (eds.), *Digital Transformation in Business and Society*, https://doi.org/10.1007/978-3-030-08277-2_3

In the present scenario, one of the major challenges facing organizations is the reconstruction of knowledge from one form to another. To successfully navigate this challenge, organizations are required to take consistent and strident steps to attain their identified goals through the efficient use of tacit knowledge. This leads to the creation of a contextual relationship between big data analytics and the management of the explicit knowledge (Davenport & Patil, 2012; Khan & Vorley, 2017). It is in this context, Khan and Vorley (2017) have identified the seminal role that big data analytics plays in capturing, acquiring, and sharing large volumes of explicit knowledge and how they may be interpreted through tacit insight. It has thus been proposed that big data and KM have a similar objective of disseminating knowledge and producing conclusive outcomes for organizations.

Technology-Driven Business Transformation

Technology has helped transform business processes and models over the years. Enterprise resource planning (ERP) and customer relationship management (CRM) helped to digitize organizational data and streamline flow of information, the new technologies like cloud computing, innovative, and cost-effective communication systems have made it possible even for small and medium enterprises (SMEs) to join the technology highway. As a result, the entire value chain has been digitized.

All the links in the chain—suppliers, distributors, retailers—were now in position to stay connected by means of these technologies. This led to a streamlined delivery process and higher efficiency of the entire system. As more and more organizations joined this technology highway, the cost of tech products decreased, starting a virtuous cycle of high demand and low price. The result was adoption of these technologies in the mainstream across the entire ecosystem.

Another simultaneous change was in progress—customer technology was being commoditized very rapidly. Widespread use of smartphones, e-commerce platforms, and availability of mobile internet were some of the major contributors in bringing technology closer to the end consumer. This new generation of consumer was not only shopping online, they were also pretty open about voicing their choices and opinions and sharing feedback on social media platforms. This information came directly from the horse mouth hence was not only true and reliable but also instantaneous. The intersection of digital technology across all ends of the spectrum is generating colossal data every minute, and this data, if processed and analyzed, can provide extremely useful information for all stakeholders.

Advantages:

1. Transformation of business models
2. Generate new sources of revenue
3. Resource optimization
4. Maximize shareholder wealth

Challenges:

1. A centralized approach for identifying and analyzing big data may not be present—though there are talks about it, not many organizations have the process and technology to identify, capture, and analyze the data coming in from various sources.
2. Talent is another limiting factor. Capable data scientists with knowledge about the industry/firm ecosystem may not be available.
3. Change needed in strategic intent—using knowledge assets to build practices around insights drawn from data is required.

The knowledge pyramid of conversion of data–information–knowledge–wisdom holds immense opportunity for organizations, countries, and the world at large. The emergence of new technologies, proliferation of the internet and smartphones, and dominance of the industry by tech giants like Apple, Amazon, Facebook, etc. has produced humungous amount of data. This has made it imperative for the field of knowledge management to develop new ways and means of analyzing and utilizing this information for global transformation.

This revolution is not just restricted to technology—it has made rapid advances to areas like manufacturing, warehousing, and logistics. Professionals of the area believe that the emergence of Industry 4.0 will be ruled by the following four:

- data volume
- computational power and connectivity
- business intelligence and analysis capabilities
- human machine interaction advances like touch screen, voice recognition, and augmented reality

Knowledge Management

The study of KM is close to touching a 25-year mark. With initial talks about knowledge in organizations in early 1990s, "knowledge creation company" by Nonaka and Takeuchi (1995)) provided a solid ground for KM to be established as a discipline in the industry. The concept garnered a lot of attention, and numerous research papers have been published in the area since.

The literature on KM speaks about these two views of a firm—resource-based view (RBV) (Barney, 1991; Penrose, 1959; Wernerfelt, 1984) and knowledge-based view (KBV) (Grant, 1996; Kogut & Zander, 1992; Spender, 1996; Teece, 1998). While the RBV focuses on the tangible and intangible resources an organization possess, the KBV takes stock of knowledge that exists within an organization and propounds that this knowledge is the source of unique and sustainable advantage the firm enjoys in the market. Both these views are based on the premise that an organization's performance is dependent more on its internal factors and less on its product positioning in the market place.

KM and intellectual capital (IC) together identify and manage knowledge assets to provide a competitive edge to a firm. IC clearly demarcates between human capital, structural capital, and relational capital (Bontis, 1999). Human capital refers to job-related skill set, structural capital refers to knowledge that exists within the organization, and relational capital exists by virtue of relations with customers, suppliers, etc.

The study of KM focuses on:

1. Tacit/explicit knowledge—based on easier or harder to capture/share (Nonaka & Takeuchi, 1995)
2. Complexity and stickiness (Kogut & Zander, 1992; McEvily & Chakravarthy, 2002)
3. Tools and techniques employed for KM like communities of practice (COPs)
4. Organizational factors like culture, trust, infrastructure, and so on, which influence the existence and smooth flow of KM processes in the organization.

Choice of the correct KM approach is a complex process and varies with a firm's circumstances.

KM refers to the process of identification of collective knowledge that exists in an organization and leveraging it to help the organization compete in the market place (Von Krogh, 1998).

From the above, it is clear that KM and strategy must work hand in hand to achieve the desired results. Linkage between the two has been suggested earlier (Hansen, Nohria, & Tierney, 1999; Hunter, Beaumont, & Lee, 2002; McDermott & O'dell, 2001; Pan & Scarbrough, 1999). KM in an organization should link to and flow from the organization's overall strategy. Researchers have emphasized on enablers, processes, and organizational performance as part of KM strategy (Lee & Choi, 2003).

KM processes typically refer to basic operations of knowledge—creation, sharing, storage, dissemination, and usage; KM enablers provide the right environment, support, and infrastructure for these processes to exist in the organization.

It is generally recognized that KM is cross-functional and touches various aspects of an organization. Many researchers have worked on this integration between different links of KM enablers, KM processes, and organizational performance (Andreeva & Kianto, 2012; Darroch, 2005; Gloet & Terziovski, 2004; Gold, Malhotra, & Segars, 2001; Hislop, 2003; Kianto, 2011; Lee & Choi, 2003; Palacios Marqués & José Garrigós Simón, 2006; Singh, Shankar, Narain, & Kumar, 2006; Zack, 2009).

Big Data and the 3V Model

In 2001, Doug Laney, an analyst at Garter (formerly META), put forth a preliminary definition of big data using a 3V model, namely, volume, velocity, and variety. Almost a decade later, in 2011, the International Data Corporation (IDC) proposed another definition for big data, stating that "big data technologies describe a new generation of technologies and architectures, designed to economically extract value from very large volume of a wide variety of data, by enabling the high-velocity capture, discovery, and/or analysis" (Sumbal, Tsui, & See-to, 2017). In recent years, much speculation and interest has surrounded big data, largely due to the availability of colossal amounts of data. This has been made possible due to the widespread dispensation of information through electronic devices and rising innovation in the field of IT.

Big data has been identified as having the potential to revolutionize the large amount of data available in present times (Davenport & Patil, 2012).

This large volume of available data has opened business organizations to the realization that there is a critical need to manage this data efficiently by storing and using it for improving accuracy in decision-making (George et al., 2014). While traditionally businesses have been dependent on the transactional data stored in relational databases for making key decisions, critical data can, however, serve as a potential reservoir for non-traditional, less-structured data forms in the present times such as emails, weblogs, social media, photographs, sensors, etc. that can unearth useful information. What makes this data even more accessible is the declining cost of storage as well as computer power. Consequently, companies everywhere are leaning toward incorporating unconventional yet potentially very valuable data, or big data, alongside the traditional enterprise data that was hitherto being used in their business intelligence analyses (Rajpathak & Narsingpurkar, 2013).

McAfee, Brynjolfsson, Davenport, Patil, and Barton (2012) observe that big data creates a potential for basing decisions on factual information as opposed to instincts, which can serve companies advantageously by providing them with a competitive edge. The vast amount of data available in present times cannot be processed adequately by the existing and predominantly traditional technologies, and it is here that big data steps in as a superior tool to extricate knowledge from this data (Kabir & Carayannis, 2013). This has led to a realization within businesses and organizations that big data holds tremendous scope in enhancing the competitive edge by serving as a treasure trove of novel ideas, valuable insights, and new knowledge.

However, it needs to be underscored that the true potential of big data can be achieved only when it is used in decision-making processes. But it is significant to note that the conventional approach of data analysis practices cannot be applied to big data. It is here that the 3V approach to big data—Variety, Velocity, and Volume—becomes significant (Kabir & Carayannis, 2013). Each of these aspects is explained as follows:

- *Variety*: The proliferation of IT has caused a significant increase in the sheer number of sources for data in present times, which in turn has led to an immense variety in the data available. Further, these resources are classed into three distinct categories, viz., structured, semi-structured, and unstructured. Structured data is that which is stored in a secure file and can be readily accessed. Data collated from vague sources and is difficult to analyze is unstructured data. Semi-structured data is that which is not organized in any special format but nevertheless contains some tags or identifiers which make it easier to process as compared to raw unprocessed data.

- *Volume*: The volume of data available is a challenge for organizations as the amount may run into terabytes and petabytes, making it difficult for the prevailing techniques of data analysis to process them.
- *Velocity*: Another challenge nowadays is the speed at which the data is generated and analyzed. Faster speeds of analysis are of utmost importance to organizations, as it will enable them to implement this processed data for better decision-making for the benefit of the organization.

The Contextual Relationship Between Big Data and Knowledge Management

Knowledge has been acknowledged as a fundamental aspect of any theorization on big data (Pauleen & Wang, 2017). Big data and analytics will fail to achieve the desired results in the absence of knowledge, as it is the latter that has led to the expansion and identification of the true potential of big data. Collection and analysis of data through the use of the appropriate techniques is dependent on the experience of the humans. It is human knowledge which will determine how the information extracted from big data is used and whether it will be applied in operational, tactical, or strategic domains. Consequently, the impact of knowledge on the role that big data will play in different sectors cannot be underestimated. This is further elucidated through an analysis of the major objectives of KM and the role that big data plays in their realization.

One of the key goals of KM is effective decision-making, which is currently accomplished by using the knowledge generated from big data analytics. This also serves to evidence the relationship between KM and big data (Crane & Self, 2014; Sumbal et al., 2017). As big data has been viewed as a potential source for generating new knowledge, it is a challenge for organizations to identify and positively exploit the true potential of digital transformation. For this, it is important to undertake a cohesive strategy to set up effective communication systems and processes to facilitate the exchange of information and data analytics, which will create a holistic approach toward KM in the digital context and foster a culture of data-driven decision-making within organizations.

Another goal of the KM system is the effective codification and sharing of the key knowledge assets. Unlike in the case of the traditional KM system, big data and the information herewith can be a source of new knowledge. That both KM and big data are concerned with some intangible asset in the form of data, information, knowledge or intelligence has also led to the development of what has been perceived as a "natural connection" between the two (Erickson & Rothberg, 2014). To ensure that these resources are efficiently utilized and meet the advanced requirements of the organizations, it has been recommended that advanced KM systems replace the erstwhile systems of rudimentary text mining tools, document analysis mechanisms, and knowledge-sharing systems (Intezari & Gressel, 2017). To generate higher value for the existing KM systems of organizations, these advanced KM systems will need to assimilate big data into their knowledge databases.

Big data analytics also becomes of crucial importance in the big picture, playing a vital role in capturing and sharing explicit knowledge by furthering its dissemination via tacit cognizance using big data analytics. In decision-making within organizations, this leads to gaining new insights into knowledge for effective and fastidious actions (Khan & Vorley, 2017).

The model presented below underlines projecting light on the contextual relationship between big data and knowledge management. At the base of the KM pyramid lies data that is composed of discrete facts and when viewed from the perspective of explicit knowledge, it refers to the movement of actions and processes in any organizational system, without any context. However, when the data is viewed in terms of a frame of reference or context, it acquires the form of information, often as metadata. When this information is further transformed through intuition, reflective thinking, personal experience, and learning of the user, it is converted into knowledge. In contrast, big data refers to the structured, unstructured, or semi-structured data collated from different sources. The application of analytics to large amounts of operational and transactional data, or explicit data, by relating it to its context gives rise to meaningful and valuable information, which directs the user toward the complete picture by studying and organizing the underlying trends and patterns revealed from the information (Fig. 3.1).

Clearly, information is a core aspect of knowledge. When meaningful information is supplied to business intelligence tools, it leads to the creation of actionable knowledge, or in other words, KM.

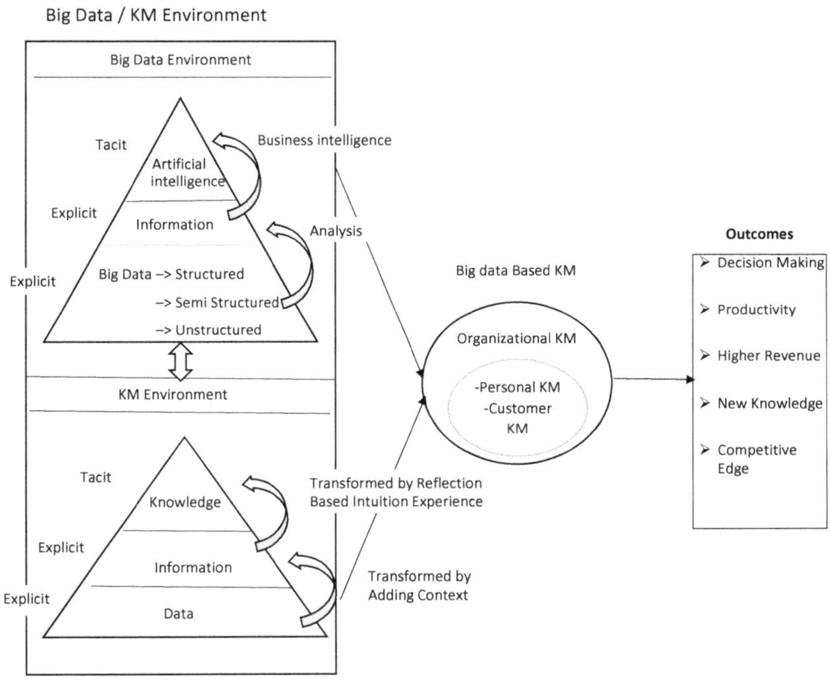

Fig. 3.1 Intersection of big data and KM in business

According to the Socialization, Externalization, Combination, and Internalization (SECI) model proposed by Nonaka, knowledge can be distinguished into two types: explicit knowledge and tacit knowledge. The former, also known as digital knowledge, is the formal and codified form of knowledge, which is captured in databases and archives and is similar to big data. On the other hand, tacit knowledge is that which exists in the minds of people and is hence much more difficult to codify. For any organization, it is the tacit knowledge of experienced employees which is significant for decision-making using big data. By bringing together the tacit knowledge of employees with the explicit knowledge that is obtained from big data, the creation of actionable knowledge and innovation is stimulated (Sumbal et al., 2017).

Clearly, big data in the form of digital data generates explicit knowledge. However, this occurs only when the available data is managed, analyzed, and embedded in the proper context. This is done by individuals, thus making tacit knowledge a pivotal aspect of big data management (Kudyba, 2014). To this has been added the generation of predictive knowledge from big data as an ancillary goal for KM decision-making (Sumbal et al., 2017).

Big data analytics has tremendous potential for organizations, and this can be realized only by leveraging it to boost decision-making (Gandomi & Haider, 2015; Sasson, Ravid, & Pliskin, 2017) Additionally, both KM and big data were conceptualized as agents that would provide deeper insights and promote better decision-making, improve productivity, generate higher revenue, and provide a competitive edge to organizations (Elgendy & Elragal, 2014; Ohlhorst, 2012; Pauleen, 2017). The conception of novel information and knowledge or explicit knowledge from big data and its implementation in decision-making alongside the use of tacit knowledge can be extremely beneficial for enhancing the accuracy of decision-making in organizations. By focusing on such aspects as the storage and dissemination of knowledge, a more informed insight may be obtained into when and how big data can serve organizations better. At the same time, a study of variables such as the nature of knowledge (tacit or explicit), can be useful for predicting the types of knowledge apt for use in different situations and industries. This will also enable the users to gain a better understanding of when and where further inputs of big data may be advantageous (Erickson & Rothberg, 2014). Analytics can also serve useful for effective KM, leading to further improvement in the functioning of organizations, thus promoting a stronger case for the intersection of big data and KM (Sumbal et al., 2017).

Role of KM in Big Data

Role of Big Data in Personal Knowledge Management (PKM)

The manner in which individuals build knowledge by utilizing such knowledge management tools which involve the collection, absorption, and the use of knowledge and the process of innovatively using knowledge has been referred to as personal knowledge management (PKM). In PKM, the focus is on the individual and his/her quest to imbibe more information, enhance work efficiency, and socialize. One of the key attributes of PKM is that it provides individuals with an opportunity to improve how they manage knowledge processes and interaction and collaborate to exchange knowledge and information with others (Huang, Pauleen, Scahill, & Taskin, 2018). Razmerita, Kirchner, and Sudzina (2009) have determined that the goal of PKM is to facilitate the work of individual knowledge workers instead of aiding the formation of organizations. Through this, PKM encourages the creation of personal knowledge for individuals by helping them discover and value information as a significant variable.

In recent times, with the application IT tools across sectors, several studies have sought to evaluate PKM in the context of the internet and Web 2.0, especially in terms of their use and management of personal knowledge (Rao & Chen, 2013; Razmerita et al., 2009) still PKM is an under-researched area (Pauleen, 2009). One of the core components of PKM is technology that enhances the efficiency and effectiveness of individual workers. This technology is useful for classifying ideas and information, and for archiving communication such as emails and other items so that they are easily accessible. In PKM, technology tools are considered to be imperative for enhancing the efficiency and confidence of the users as well as for the development of more effective information and knowledge management skills and dissemination and receipt of up-to-date information (Agnihotri & Troutt, 2009; Benitez & Pauleen, 2009). Liu, Wang, and Lin (2017) in their study on the applications of big data in PKM concluded that five domains:

1. *Time management*: Increased information sharing through facilitation of provision of resources which were earlier not available like software for converting the earlier not-so-easy convertible files to now free availability of such resources
2. *Computer usage efficiency management*: Retrieval of contextual information while using computers and internet gets facilitated as an individual's behavior is understood and reciprocated to through the use of Big Data.
3. *Mobile devices using behavioral management*: Behavioral parts of individuals can be analyzed using mobile phones without time and space constraints, this data is of massive use to organizations and individuals.
4. *Healthcare management*: Individuals users of mobile technology have access to unlimited apps and software, which may be of help in managing personal health by providing timely updates and reminders

about food, medicine, walk, etc. Technology has gone to the extent of being able to provide even ECG of the wearer, which may go a long way in avoiding cardiac problems by early diagnosis (Murdoch & Detsky, 2013).

5. *Browsing surfing management*: Interface between big data and an individual's surfing behavior can help in finding relevant information faster than usual. Tech devices can be made to remember the more frequently used, searched, visited websites/places hence improving the surfing experience.

The Role of Big Data in Customer Knowledge Management(CKM)

One of the key assets for any organization is customer knowledge, and hence, customer knowledge management or CKM is an important aspect for business. CKM facilitates the exchange of customer knowledge both within firms and between customers and firms in order to facilitate learning from, about, and with customers (Gibbert, Leibold, & Probst, 2002; Martín-de Castro, 2015). When customers are involved in the firm's practices through CKM, the external environment of the firm is linked to its internal environment. This, in turn, enables the transfer and sharing of information within the firm and with customers. However, there have been identified several challenges in converting customers' tacit knowledge (experience, ideas, information, difficulties, requirements, data, etc.) to explicit knowledge (useful ideas for resolving the problems of customers, for innovative services, or for upgrading current services (Taherparvar, Esmaeilpour, & Dostar, 2014).

Major retail companies such as Amazon.com, eBay, Walmart, and others have been able to successfully manage large amounts of knowledge and communicate with customers through the inclusion of real-time KM for the generation of knowledge using big data and analysis techniques (Davenport & Patil, 2012; McAfee et al., 2012; Waller & Fawcett, 2013; Sumbal et al., 2017). In these companies, the data accumulated from different sources is converted into business intelligence and knowledge via analytics. Big data analytics is also crucial in the assessment of the behavioral patterns of consumers, which is accomplished using such technology as machine-to-machine platforms and social media. The main objective here is to encapsulate the structured and unstructured data through big data in a manner that will transform it into actionable insights by using CKM and lead to the generation, sharing, and storage of knowledge (Chan, 2014). This integration of big data and CKM for the generation of actionable insights for businesses support the creation of value for organizations.

Integration of Knowledge Management and Big Data

While both these systems are being implemented in activities like consumer survey, performance appraisal, capacity enlargement, etc., it can be observed that the independent framework of these two technologies share association with each other at multiple levels with KM playing the foundational role because of its earlier inception and time-tested performance in assisting with the installation of big data systems. However, big data is slowly adding benefits to KM in exponential ways that will make it a productive necessity for the future prospects of KM. Figure 3.2 presents the integration of the two. This integration is based on the premise that knowledge is central to all meaningful human activity. While analyzing big data, use of knowledge management is imperative as the problem or opportunity analysis must be clearly stated and form the base of the entire exercise. In the former case, big data analytics may be utilized to plan for agricultural management through geographical information or planning a city through the use of traffic patterns and trends. Opportunity identification may be purpose at hand, Google cloud platforms and Amazon web services help in the same. Organizations IT infrastructure, collaborative technologies act as the backbone for such analysis. Knowledge possessed by the employees of an organization (Tacit) and easily available coded/explicit knowledge helps us enter into the realm of big data possessed with the right weapons. Big data sources such as media, social networking sites, sensor data, archives, and log data provide humongous amount of raw data, which can be processed to find meaningful trends and opportunities by identifying and classifying the useful from the rest.

The process yields extremely useful actionable insights for managers who are in a better position to make informed decisions now. This in turn results in various benefits for the organization in terms of economic value, supply chain efficiency, and innovation.

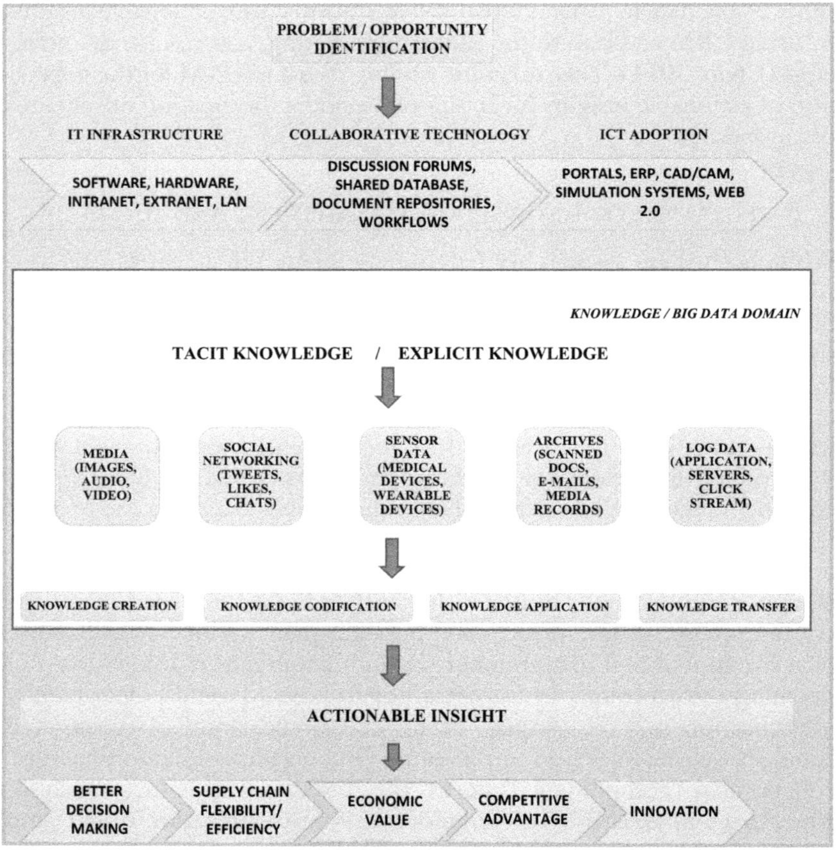

Fig. 3.2 Knowledge management—harnessing the power of big data

Therefore, the interaction of two different yet associated practices paved way to the illustrated model that can provide new business value and faster return on economic investments, enhance business operations through better decision-making, incorporate trust, promote personal and organizational-level learning, and ultimately the perks of a hybrid system that is adaptable to future needs is what comes as a bonus.

Conclusion

The twenty-first century has witnessed a major transition toward a knowledge economy of unlimited scope. During the past few years, the world has rapidly grown to connect products and services that were earlier perceived as impossible and all this can be attributed to the rising technological advancements of the second millennium. Along the same timeline, the inclusion of KM as a pivotal partner in productive practices inside companies and organizations have taken a huge leap to withstand the competition. The inevitable use of big data in a world that is run by networking has also penetrated into KM and its future, making it prone to theoretical and practical evolution on a continuous basis.

Digital transformation is inevitable in the current times. The need of the hour is that organizations understand and realize the importance of potential learning opportunities that may result from the analysis of big data. Organizations with a futuristic leadership have already woken up to the fact that "digital" is the new mantra. They are in the process of taking a stock of their current capabilities and working on a digital agenda to reskill their employees and introduce them to new learning environments and opportunities. They are in full recognition of the myriad changes' customer expectation has undergone during the last decade and appreciate the fact that some major changes will be required to meet them.

The traditional organization trained and updated their employee base by providing technical and managerial skill set training programs from time to time. While recognition of gaps in technical skills could be done using skill matrix and competence models, and appropriate training could be provided; managerial skills could be honed through communication, negotiation, and decision-making programs.

These programs were effective and yielded the desired results until the rate of change in business was incremental and the conventional theories of economics applied on the markets. With the current disruptive changes occurring in business environment across the globe, the need of the hour is development of short, situation-specific, on-the-go solutions and suggestions being provided to the employee, which may be used and referred to at their convenience of time and space.

Organizations' collective knowledge must be digitized and prioritized according to the current demands of the market. Knowledge management needs to be innovative and creative in its approach to adapt and realign itself to the changing information needs of organizations. Setting up centers to leverage knowledge assets of the organization, data driven insights which aid decision making and alignment of strategy with business objectives will lead to success of knowledge management.

Hence, the quantitative and real-world analysis of the meeting of a robust KM with an emerging big data technology forms the framework of this study. By identifying the correlation between these big data and KM, a connection that links both together could be discovered that shed light on its different functionalities. The results of this convergence for a sole entity can be termed as personal knowledge management (PKM), expanding into customer knowledge management (CKM) and finally into organizational knowledge management (OKM) (Pee & Kankanhalli, 2009).

The same convergence also results in the advent of a new corporate methodology, which is increasingly being used in all forms of organizations; big and small for informed decision-making. This is basically the enhanced version of decision-making, which is infused by leveraging the added potential of existing KM practices and new big data capabilities, but the effectiveness of this intersection lies on how well these two practices mingle and co-exist in a corporate's workflow. It is this co-existence that has to be closely moderated for successful implementation and enhancement of KM by featuring big data as the prime contributor.

Deliberate, timely, and careful alignment of knowledge management with big data is the key to success in the current business scenario.

Limitation and Future Scope

Although big data is gaining attention in research but limited research exists putting big data and KM in one framework; hence, it remains underdeveloped within the KM literature. This chapter is a conceptual study that need empirical testing. The literatures found to the subject in discussion depicted association between big data and only few variables of knowledge management but not all areas have been thoroughly studied. The prevalent studies have only considered a few sectors to demonstrate the association between big data and knowledge management. Thus, there is still a huge scope of testing this linkage empirically in other sectors to understand its effectiveness on other industries. The review of the literature studied between big data and various areas of knowledge management, but its linkage in a broader framework still needs to be empirically tested, which opens up the scope for future research.

Appendix

Linkage between the variables

Title of the research paper	*Main findings*	*Source of literature*
Interrelationship Between Big Data and Knowledge Management: An Exploratory Study in the Oil and Gas Sector	• Generation of predictive knowledge • Improve KM capabilities • Effective decision-making	Sumbal et al., *Journal of Knowledge Management* (2017)
Big Data Systems: Knowledge Transfer or Intelligence Insights?	• Similarity between big data and KM structure	Rothberg and Erickson, *Journal of Knowledge Management* (2017)
Creation of Knowledge-Added Concept Maps: Time Augmentation via Pairwise Temporal Analysis	• High correlation between temporal assessment and subjective assessments	Sasson et al., *Journal of Knowledge Management* (2017)
Cognitive Big Data: Survey and Review on Big Data Research and Its Implications. What's Really New in Big Data?	• Five-trait framework for big data 1. Socio-technical system 2. Data space 3. Data richness 4. Knowledge management 5. Decision-making 6. Sensory/visualization presentation	Lugmayr, Stockleben, Scheib, and Mailaparampil, *Journal of Knowledge Management* (2017)
Facilitating Knowledge Management through Filtered Big Data: SME Competitiveness in an Agri-Food Sector	• Big data consumer analytics helps in tapping explicit and tacit knowledge • Effective implementation of KM processes needed in SMEs	O'Connor and Kelly, *Journal of Knowledge Management* (2017)
Big Data Text Analytics: An Enabler of Knowledge Management	• Empower KM	Khan and Vorley, *Journal of Knowledge Management* (2017)
Big Data and Knowledge Management; A Case of DÉJÀ VU or Back to the Future?	• Pattern analysis and prediction with application of algorithms • Data-driven decision-making • Big data as a contributor to KM	Tian, *Journal of Knowledge Management* (2017)
Information and Reformation in KM Systems: Big Data and Strategic Decision-Making	• Identified four types of data based decisions which aids KM systems in handling big data and advanced analytics	Intezari and Gressel, *Journal of Knowledge Management* (2017)

(*continued*)

(continued)

Title of the research paper	*Main findings*	*Source of literature*
The Concept of Big Data Applied in Personal Knowledge Management	• Five areas of personal knowledge management with application of big data: 1. Time management 2. Computer usage efficiency management 3. Mobile device behavior management 4. Health management 5. Browser surfing management	Liu et al., *Journal of Knowledge Management* (2017)
Big Data and Knowledge Management: Establishing a Conceptual Foundation	• Connection between KM, IC and application of big data and business analytics exists • With advance use of big data knowledge assets can be beneficial	Erickson and Rothberg, *Electronic Journal of Knowledge Management* (2014)
Big Data, Tacit Knowledge and Organizational Competitiveness	• Inhabited knowledge in big data has potential of creating economic value • Major source of competitive advantage	Kabir and Carayannis, *ICICKM* (2013)
Big Data Customer Knowledge Management	• Integration of CRM, CKM and big data as value creation strategies • Increase quality • Improves trustworthiness	Chan, *Communications of the IIMA* (2014).

References

Agnihotri, R., & Troutt, M. D. (2009). The effective use of technology in personal knowledge management: A framework of skills, tools and user context. *Online Information Review, 33*(2), 329–342.

Andreeva, T., & Kianto, A. (2012). Does knowledge management really matter? Linking knowledge management practices, competitiveness and economic performance. *Journal of Knowledge Management, 16*(4), 617–636.

Barney, J. (1991). Firm resources and sustained competitive advantage. *Journal of Management, 17*(1), 99–120.

Benitez, E. A., & Pauleen, D. (2009). Brainfiltering: The missing link between PKM and PIM? *AMCIS 2009 Proceedings*, 13.

Bontis, N. (1999, August). Managing an organizational learning system by aligning stocks and flows of knowledge. In *Academy of Management Proceedings* (Vol. 1, pp. J1–J2). Briarcliff Manor, NY: Academy of Management.

Chan, J. O. (2014). Big data customer knowledge management. *Communications of the IIMA, 14*(3), 5.

Crane, L., & Self, R. J. (2014, September). Big Data Analytics: A threat or an opportunity for Knowledge Management? In *International Conference on Knowledge Management in Organizations* (pp. 25–34). Cham: Springer.

Darroch, J. (2005). Knowledge management, innovation and firm performance. *Journal of Knowledge Management, 9*(3), 101–115.

Davenport, T. H., & Patil, D. J. (2012). Data scientist. *Harvard Business Review, 90*(5), 70–76.

Elgendy, N., & Elragal, A. (2014, July). Big data analytics: A literature review paper. In *Industrial Conference on Data Mining* (pp. 214–227). Cham: Springer.

Erickson, S., & Rothberg, H. (2014). Big data and knowledge management: Establishing a conceptual foundation. *Electronic Journal of Knowledge Management, 12*(2), 101.

Gandomi, A., & Haider, M. (2015). Beyond the hype: Big data concepts, methods, and analytics. *International Journal of Information Management, 35*(2), 137–144.

George, G., Haas, M. R., & Pentland, A. (2014). Big Data and management. *Academy of Management Journal, 57*(2), 321–326.

Gibbert, M., Leibold, M., & Probst, G. (2002). Five styles of customer knowledge management, and how smart companies use them to create value. *European Management Journal, 20*(5), 459–469.

Gloet, M., & Terziovski, M. (2004). Exploring the relationship between knowledge management practices and innovation performance. *Journal of Manufacturing Technology Management, 15*(5), 402–409.

Gold, A. H., Malhotra, A., & Segars, A. H. (2001). Knowledge management: An organizational capabilities perspective. *Journal of Management Information Systems, 18*(1), 185–214.

Grant, R. M. (1996). Toward a knowledge-based theory of the firm. *Strategic Management Journal, 17*(S2), 109–122.

Hansen, M. T., Nohria, N., & Tierney, T. (1999). What's your strategy for managing knowledge. *The Knowledge Management Yearbook 2000–2001, 77*(2), 106 –116.

Hislop, D. (2003). Linking human resource management and knowledge management via commitment: A review and research agenda. *Employee Relations, 25*(2), 182–202.

Huang, Y. M., Pauleen, D., Scahill, S., & Taskin, N. (2018). A PKM–based decision-making training program for personal healthcare: An action learning approach. *International Journal of Knowledge Management, 14*(3), 101–114.

Hunter, L., Beaumont, P., & Lee, M. (2002). Knowledge management practice in Scottish law firms. *Human Resource Management Journal, 12*(2), 4–21.

Intezari, A., & Gressel, S. (2017). Information and reformation in KM systems: Big data and strategic decision-making. *Journal of Knowledge Management, 21*(1), 71–91.

Kabir, N., & Carayannis, E. (2013, January). Big data, tacit knowledge and organizational competitiveness. In *Proceedings of the 10th International Conference on Intellectual Capital, Knowledge Management and Organisational Learning: ICICKM* (p. 220).

Khan, Z., & Vorley, T. (2017). Big data text analytics: An enabler of knowledge management. *Journal of Knowledge Management, 21*(1), 18–34.

Kianto, A. (2011). The influence of knowledge management on continuous innovation. *International Journal of Technology Management, 55*(1/2), 110–121.

Kogut, B., & Zander, U. (1992). Knowledge of the firm, combinative capabilities, and the replication of technology. *Organization Science, 3*(3), 383–397.

Kudyba, S. (2014). Information creation through analytics. In *Big data, mining, and analytics. Components of strategic decision making* (pp. 17–48). Boca Raton: CRC Press, Taylor & Francis Group.

Lee, H., & Choi, B. (2003). Knowledge management enablers, processes, and organizational performance: An integrative view and empirical examination. *Journal of Management Information Systems, 20*(1), 179–228.

Liu, C. H., Wang, J. S., & Lin, C. W. (2017). The concepts of big data applied in personal knowledge management. *Journal of Knowledge Management, 21*(1), 213–230.

Lugmayr, A., Stockleben, B., Scheib, C., & Mailaparampil, M. A. (2017). Cognitive big data: Survey and review on big data research and its implications. What is really "new" in big data? *Journal of Knowledge Management, 21*(1), 197–212.

Martín-de Castro, G. (2015). Knowledge management and innovation in knowledge-based and high-tech industrial markets: The role of openness and absorptive capacity. *Industrial Marketing Management, 47*, 143–146.

McAfee, A., Brynjolfsson, E., Davenport, T. H., Patil, D. J., & Barton, D. (2012). Big data: The management revolution. *Harvard Business Review, 90*(10), 60–68.

McDermott, R., & O'dell, C. (2001). Overcoming cultural barriers to sharing knowledge. *Journal of Knowledge Management, 5*(1), 76–85.

McEvily, S. K., & Chakravarthy, B. (2002). The persistence of knowledge-based advantage: An empirical test for product performance and technological knowledge. *Strategic Management Journal, 23*(4), 285–305.

Murdoch, T. B., & Detsky, A. S. (2013). The inevitable application of big data to health care. *JAMA, 309*(13), 1351–1352.

Nonaka, I., & Takeuchi, H. (1995). *The knowledge creation company: How Japanese companies create the dynamics of innovation*. New York: Oxford University Press.

O'Connor, C., & Kelly, S. (2017). Facilitating knowledge management through filtered big data: SME competitiveness in an agri-food sector. *Journal of Knowledge Management, 21*(1), 156–179.

Ohlhorst, F. J. (2012). *Big data analytics: Turning big data into big money*. John Wiley & Sons.

Palacios Marqués, D., & José Garrigós Simón, F. (2006). The effect of knowledge management practices on firm performance. *Journal of Knowledge Management, 10*(3), 143–156.

Pan, S. L., & Scarbrough, H. (1999). Knowledge management in practice: An exploratory case study. *Technology Analysis & Strategic Management, 11*(3), 359–374.

Pauleen, D. (2009). Personal knowledge management: Putting the "person" back into the knowledge equation. *Online Information Review, 33*(2), 221–224.

Pauleen, D. J. (2017). Davenport and Prusak on KM and big data/analytics: Interview with David J. Pauleen. *Journal of Knowledge Management, 21*(1), 7–11.

Pauleen, D. J., & Wang, W. Y. (2017). Does big data mean big knowledge? KM perspectives on big data and analytics. *Journal of Knowledge Management, 21*(1), 1–6.

Pee, L. G., & Kankanhalli, A. (2009). A model of organisational knowledge management maturity based on people, process, and technology. *Journal of Information & Knowledge Management, 8*(02), 79–99.

Penrose, E. T. (1959). *The theory of the growth of the firm*. New York: John Wiley.

Rajpathak, T., & Narsingpurkar, A. (2013). Managing knowledge from Big Data analytics in product development. White Paper, Tata Consultancy Services.

Rao, Y., & Chen, S. (2013, June). Research on KAAS-based social knowledge collaboration service mechanism and application. In *The Proceeding of 2013 International Conference on Artificial Intelligence and Software Engineering*.

Razmerita, L., Kirchner, K., & Sudzina, F. (2009). Personal knowledge management: The role of Web 2.0 tools for managing knowledge at individual and organisational levels. *Online Information Review, 33*(6), 1021–1039.

Rothberg, H. N., & Erickson, G. S. (2017). Big data systems: Knowledge transfer or intelligence insights? *Journal of Knowledge Management, 21*(1), 92–112.

Sasson, E., Ravid, G., & Pliskin, N. (2017). Creation of knowledge-added concept maps: Time augmention via pairwise temporal analysis. *Journal of Knowledge Management, 21*(1), 132–155.

Singh, M. D., Shankar, R., Narain, R., & Kumar, A. (2006). Survey of knowledge management practices in Indian manufacturing industries. *Journal of Knowledge Management, 10*(6), 110–128.

Spender, J. C. (1996). Making knowledge the basis of a dynamic theory of the firm. *Strategic Management Journal, 17*(S2), 45–62.

Sumbal, M. S., Tsui, E., & See-to, E. W. (2017). Interrelationship between big data and knowledge management: An exploratory study in the oil and gas sector. *Journal of Knowledge Management, 21*(1), 180–196.

Taherparvar, N., Esmaeilpour, R., & Dostar, M. (2014). Customer knowledge management, innovation capability and business performance: A case study of the banking industry. *Journal of Knowledge Management, 18*(3), 591–610.

Teece, D. J. (1998). Capturing value from knowledge assets: The new economy, markets for know-how, and intangible assets. *California Management Review, 40*(3), 55–79.

Tian, X. (2017). Big data and knowledge management: A case of déjà vu or back to the future? *Journal of Knowledge Management, 21*(1), 113–131.

Von Krogh, G. (1998). Care in knowledge creation. *California Management Review, 40*(3), 133–153.

Waller, M. A., & Fawcett, S. E. (2013). Data science, predictive analytics, and big data: A revolution that will transform supply chain design and management. *Journal of Business Logistics, 34*(2), 77–84.

Wernerfelt, B. (1984). A resource-based view of the firm. *Strategic Management Journal, 5*(2), 171–180.

Zack, M. H. (2009). *Knowledge and strategy*. Routledge.

CHAPTER 4

Social Media in Marketing Communications: A Synthesis of Successful Strategies for the Digital Generation

Ogechi Adeola, Robert Ebo Hinson, and Olaniyi Evans

Introduction

Numerous studies have investigated social media in marketing communications and have advanced our understanding of how it unfolds (e.g., Berthon, Pitt, Plangger, & Shapiro, 2012; Hudson & Thal, 2013; Kumar, Choi, & Greene, 2016). We currently have a mounting body of (1) evidence which shows that social media influences marketing outcomes (e.g., Agnihotri, Dingus, Hu, & Krush, 2016; Chung & Austria, 2010; Erdoğmuş & Cicek, 2012), and (2) several frameworks that offer an explanation of the complex social media marketing process (e.g., Effing & Spil, 2016; Felix, Rauschnabel, & Hinsch, 2017; Kietzmann, Hermkens, McCarthy, & Silvestre, 2011).

O. Adeola (✉) • O. Evans
Pan-Atlantic University, Lagos, Nigeria
e-mail: oadeola@lbs.edu.ng

R. E. Hinson
University of Ghana Business School, Accra, Ghana

B. George, J. Paul (eds.), *Digital Transformation in Business and Society*, https://doi.org/10.1007/978-3-030-08277-2_4

In recognition of its critical importance, according to Gallaugher and Ransbotham (2010), many companies have established online presence and continuous conversations with their consumers (e.g., Old Spice, TEDx, Starbucks, Pampers, Xbox, Zappos, Ford Motor Company, Cisco, JetBlue, Uber). These companies have shifted relationships with consumers from dialogue to trialogue, with consumers engaging one another and the firms in meaningful relationships (Lipsman, Mudd, Rich, & Bruich, 2012; Mangold & Faulds, 2009). With the use of social media, these organisations have forged relationships with new as well as existing customers and formed virtual communities that interactively join forces to recognise problems and advance solutions (Tsimonis & Dimitriadis, 2014).

These interactions, in turn, have changed the traditional roles of both the customer and the seller in exchange relationships. The consumers, on their part, have reduced their dependence on the traditional media (i.e., television, radio, newspapers, and magazines) as a source of information to guide their purchasing decisions. They now demand more control over their media consumption and prefer on-demand and immediate access to information, at their convenience (Vollmer & Precourt, 2008). By utilising the social media to conduct their information searches and make purchasing decisions (Vollmer & Precourt, 2008), the consumers increasingly perceive social media as a more trustworthy source of information on products and services than corporate-sponsored communications (Foux, 2006). In particular, the consumers respond to this information in a manner 'that directly influence all aspects of consumer behaviour, from information acquisition to post-purchase expressions of satisfaction and dissatisfaction' (Mangold & Faulds, 2009, p. 361). They add value to the information by generating content, and they can influence purchase decisions of others in peer-to-peer interactions (Sashi, 2012).

Additionally, from giants like IBM and Starbucks to the local barber, organisations have realised the fact that vast amounts of information about their products and services are communicated by individual consumers on social media. They know that social media, today, provides one of the best opportunities available to a brand for connecting with prospective consumers (Neti, 2011). In recognition of these processes, therefore, managers, who consistently wield a high level of control over company-to-consumer messages, must learn new strategies to influence discussions in the social media space. Whether it is an individual, a startup, small business, or a large corporation, successful strategies must be put in place to leverage the power of social media.

However, despite the recognition that social media is effective in connecting with customers (Selina & Milz, 2009; Hackworth & Kunz, 2011; Tsimonis & Dimitriadis, 2014), there is limited understanding regarding the successful strategies for social media marketing. Most of the relatively few studies in the literature consider social media as a novel marketing tool (e.g., Berinato, 2010; Kirtiş & Karahan, 2011; Bowen & Bowen, 2016), which can boost marketing communication effectiveness (e.g., Trusov, Bucklin, & Pauwels, 2009; Dholakia & Durham, 2010), and very few focus on specific strategies for social media marketing (e.g., Mangold & Faulds, 2009). To the best of our knowledge, there appears to be limited research examining the successful strategies that firms use for their corporate fan pages on social media like Twitter, Facebook, or YouTube.

Considering this gap, the increasing recognition of the significance of social media in marketing communications, its rapidly expanding evidence base, and the numerous social media marketing frameworks that are emerging, this study seeks to increase the understanding of social media in marketing communications by undertaking a conceptual synthesis of the successful strategies in literature and from business practices. The study identifies the successful strategies in social media marketing and includes specific actions related to these strategies that can be utilised to achieve success. This study, therefore, differs from other studies in the literature, as it delves into the 'how-to' of social media marketing (i.e., specific procedures and strategies). Systematically identifying these action-oriented strategies can serve as practical guidance related to specific tasks to include in the planning and/or execution of social media marketing. It is believed that the explicit focus on specific strategies that can be used to operationalise social media in marketing communications would make a significant contribution to the literature.

In the next section, we present a brief historical overview of social media; and issues related to social media marketing are described. The following section further discusses the electronic word-of-mouth (eWOM) framework. In the fourth section, the social media marketing strategies are synthesised; while the last section covers the concluding thoughts of this study.

Brief History of Social Media Marketing

In the last decade, social media has enjoyed phenomenal success. As at 2019, Facebook, a social networking website, has 2.27 billion active users worldwide; Twitter, a micro blogging website, currently hosts more than

349 million active users with about 500 million tweets per day; and YouTube, a video-sharing website, has over 3 billion daily video views (Internetlivestats.com, 2019).

The origin of social media marketing can be traced back to the 1990s when scholars began to highlight the implications of the Internet's interactivity for marketing and suggesting the possibility of IT developments rendering traditional marketing practices obsolete (van Raaij, 1998). Within a few years, the huge impacts of the web on consumer behaviour can be described as similar to those of the Industrial Revolution, of the printing press, or the railroads (Sharma & Sheth, 2004). Due to the impact of the web, consumer preferences are no longer only influenced by the traditionally defined controllable and uncontrollable stimuli, but also by the web experience, or the online atmospherics consisting of online controllable factors such as website usability, interactivity, aesthetics, and online marketing mix (Constantinides, 2009; Fotis, 2015), and by the website's interface, quality, satisfaction, and experience (Darley, Blankson, & Luethge, 2010).

The Internet had become a great enabler of consumer power, causing a steady decline in the influence of traditional mass media (Constantinides, 2009; Urban, 2003). As noted by Sharma and Sheth (2004, p. 696), 'the web is fundamentally changing, and will continue to change marketing thought and practice', social media realigned marketing focus from a supplier to a customer standpoint, aiding a customer-centric rather than a mass-market approach. Therefore, it is the customer now, and not the product, that initiates marketing activities (Fotis, 2015). In other words, the key lies with the change in the consumer's mindset, leading to 'a culture of active engagement, immediate access to anyone and for anything and constant communication. This is, of course, something that the traditional advertisement and other marketing methods could not ever achieve' (Kaushik, 2012, p. 92).

The first companies to use social media for marketing were Usernet in 1979, Bulletin Board System in 1979, Online Service in early to mid-1980s, CompuServe in 1980, AOL in 1983, Genie in 1985, Internet Relay Chat in 1988, and ICQ in mid-1990. The earliest social media were dating sites and online forums, such as Six Degrees in 1997 and Asian Avenue, MiGente, and Black Planet in 1997–2001, Live Journal in 1999 and Massive Multiplayer Online and Role Playing Games in early 2000. Then came along Friendster in 2002, LinkedIn in 2003, Myspace in 2003, Photo bucket in 2003, Delicious in 2003, Digg in 2004, Flickers in 2004,

Facebook in 2004, YouTube in 2005, Twitters in 2006, Kontain in 2008, and Posterous in 2009 (Kaushik, 2012).

With these platforms and more in the cyberspace, social media evolved into eight types of applications, namely, social networking sites, blogs, wikis, microblogs, consumer review sites, content community sites, Internet forums, and location-based social media. These applications created a virtual space within which users listen, express, share, discuss, engage, and interact through user- and brand-generated content. As social media became popular and the number of users increased, the commercial value soon became apparent. Marketers started to consider this shift in the media landscape, rewrote conventional strategies, and transformed the basic campaigning structure.

In the last decade, social media has transformed 'the web from a "one to many" to a "many to many" communication channel, and simultaneously to a "one to one" channel of interaction, allowing users not only to communicate but also to interact and engage' (Fotis, 2015, p. 34). This transformation has implications for marketing in the form of consumer empowerment, greater information transparency, possibility for product co-creation and personalisation, content generation and publishing, capacity to sanction/reward brands for product performance, and ethical and social responsibilities (Buhalis & O'Connor, 2005; Rodriguez-Ardura Meseguer-Artola, 2010).

There are also unintended implications. Social media makes consumers highly individualistic, more information-intensive, more time-driven and demanding, dictating time and mode of communication, and with bigger expectations (Akehurst, 2009). Moreover, the huge volume of information on the web leads to information overload, negatively affecting users' ability to locate information tailored to their specific needs, and causing other pathologies of information such as infobesity and information anxiety (Bawden & Robinson, 2009).

Theoretical Framework: Electronic Word-of-Mouth (eWOM)

The framework we used addresses the linkage between word-of-mouth (WOM) and social media (see Fig. 4.1). WOM can be defined as 'the act of exchanging marketing information among consumers and plays an essential role in changing consumer attitudes and behaviour towards prod-

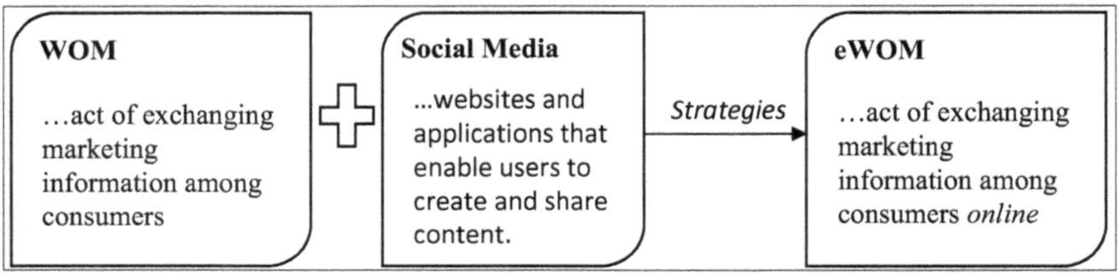

Fig. 4.1 The linkage between WOM and eWOM. Source: Authors' representation

ucts and services' (Chu & Kim, 2011, p. 47). WOM has been shown to significantly affect consumer attitudes and is seven times more effective than traditional print advertising in influencing brand-switching decisions (Engel, Kegerreis, & Blackwell, 1969; Katz & Lazarsfeld, 1955). Consumers often rely on WOM when they are in need of information to aid their purchase decisions. This is due to the fact that WOM is created and delivered by friends and family and therefore is a more trustworthy source of information than company-generated messages (Chu & Kim, 2011; Crotts 1999).

The emergence of WOM in social media has led to eWOM (electronic WOM) (see Fig. 4.1). According to Hennig-Thurau et al. (2004, p. 39), eWOM is 'any positive or negative statement made by potential, actual, or former customers about a product or company, which is made available to a multitude of people and institutions via the Internet'. eWOM occurs in a wide variety of online media, such as blogs, emails, virtual consumer communities, social networking sites, and consumer review websites and forums (Thorson & Rodgers, 2006). This eWOM acquired from friends and family is construed to 'be more credible, honest, and trustworthy than that generated from marketers, since the communicators are not compensated for the referral. Advancements in the Internet and Web 2.0 technologies now allow consumers to access personally meaningful critiques not only from friends and relatives but from strangers' (Pan & Crotts, 2012, p. 43).

In the social media marketing literature, there has been an increasing interest in understanding how eWOM affects sales, diffusion, and other marketing performance outcomes (i.e., Bruce, Foutz, & Kolsarici, 2012; Chevalier & Mayzlin, 2006; Godes & Mayzlin, 2004; Onishi & Manchanda, 2012; Trusov et al., 2009; Van den Bulte & Lilien, 2001; Villanueva, Yoo, & Hanssens, 2008). For example, Godes and Mayzlin

(2004) showed that online discussion forum activities influence television show ratings. Chevalier and Mayzlin (2006) demonstrated how user-generated online book reviews affect book sales. Trusov et al. (2009) showed that referrals to join an online social media platform affect a website's growth. Further, Onishi and Manchanda (2012) depicted that blog activity affects sales of a set of Japanese brands.

Consistent with the findings of the literature above on how eWOM affects a product's marketing performance, marketing managers combine WOM and social media to create eWOM, via effective social media marketing strategies (Fig. 4.1). Using effective strategies, the marketing managers engage the consumers in social interactions with the firm and the product and thereby get rewarded with likes, comments, or dissemination to the consumers' social connections. Consumers freely create and disseminate brand-related eWOM to friends, family, and acquaintances (Vollmer & Precourt, 2008). Through these social media interactions, consumers can show their brand preference to their connections, thus leading to eWOM (Chu & Kim, 2011). Utilising effective strategies, therefore, a firm can employ WOM on social media to generate eWOM (i.e., buzz marketing) and make its products go viral.

Successful Strategies

As more firms incorporate social media into their marketing communications, emphasis needs to be placed on creating and adopting effective strategies that can generate eWOM and make the brand go viral. The strategies outlined in this study are discrete social media practices that researchers have identified as universally beneficial for organisations to adopt. The aim here is not to generate radically new ideas for social media marketing in a discrete sense; each of the strategies is being operationalised in some organisations today. Rather, the purpose of this synthesis is to provide a measure of coherence, purpose, and success to social media marketing.

By synthesising these strategies, this study is able to aggregate the critical themes from the available literature and international experience to suggest actions that marketing practitioners can employ to ensure success. In order for a firm to achieve its goals in a social media campaign, it has to identify the successful strategies first and ensure that those strategies run in tandem with its overall marketing strategy. The strategies could serve well to nurture the creative and adaptive marketing capabilities of the organisation.

Create Good Content

Social media and content have a symbiotic relationship: without social media, the content cannot be accessible, and without great content, social media is pointless (Kaplan & Haenlein, 2011; Kietzmann et al., 2011). Studies such as Dolan (2016) have shown that social media content design and delivery influence the engagement behaviour of Facebook fans. Consumers are more likely to generate eWOM about the firm and its products when they have adequate knowledge (Mangold & Faulds, 2009). For example, Starbucks gives tips to help people use its products on its 37-million-fans Facebook page. Pages are actually a great means of connecting with audiences where they hang out.

There are three main components to social media content strategy: type of content, time of posting and frequency of posting (Chauhan & Pillai, 2013). The type of content relies on form (i.e., how the information is presented—text only, images, links, video, amongst others) and context (i.e., company voice and platform trends). Since each audience is unique, firms must identify the best type of content, time of posting, and frequency of posting in order to accrue more engagement or more likes and follows. Moreover, to get more engagement, likes and shares, firms may need to continue to inspire their audience; make their content shareable, fun and useful; use great graphics and video; and cross-promote to other channels (Berthon et al., 2012).

Get the Right People to Spread the Message

Creating successful social media marketing requires the right people to spread the message (Pfeffer, Zorbach, & Carley, 2014; Tuten & Solomon, 2014). To effectively transform an ordinary message into an eWOM phenomenon, three groups of messengers are needed: market mavens, social hubs, and salespeople. In line with the conversion theory (Moscovici, 1980), these minority group of messengers have a huge effect on the majority. According to Kaplan and Haenlein (2011, p. 256), '[c]onsistent with classical laws of concentration, 20% of messengers can be expected to carry 80% of the load; it is, therefore, especially crucial to select wisely the initial hosts' for the message.

Market mavens are individuals with access to vast marketplace information, and proactively discuss and diffuse information (Feick & Price, 1987). The market mavens first receive the message and transmit it to their immediate social hub, who begins the eWOM phenomenon. Social hubs are people with large social connections (Goldenberg, Libai, & Muller, 2001), who facilitate immediate transmission of the message to a large number of consumers. In cases where market mavens are not particularly convincing in transmitting the message, salespeople would be needed to amplify and transmit it to the immediate social hub for further dissemination (Kaplan & Haenlein, 2011).

Create Virtual Communities of Individuals with Common Interests

The literature has shown that consumers like to network with people of similar interests and desires (e.g., Mangold & Faulds, 2009; Weber, 2009). Organisations can, therefore, leverage on this desire by creating virtual communities of like-minded individuals, centred on shared interests and values (Rolls, Hansen, Jackson, & Elliott, 2016; Weber, 2009). Virtual communities encourage interaction, sometimes focusing on a particular interest or just to communicate. Users can interact over a shared passion through various means, namely, chat rooms, message boards, virtual worlds, or social networking sites. Since a virtual community potentially crosses geographical and political boundaries, it provides unprecedented leverage to grow and maintain a firm's network, as well as create eWOM.

For example, Roadrunner Records' website, www.roadrunnerrecords.com, has a forum for fans of rock and metal music where users connect and share interests. The Obama 2008 presidential campaigns leveraged on their online presence to attract supporters and provide information. Toyota leveraged on a blog written by a team of a father and a son who embark on annual 5000-mile adventures to unusual destinations in their Toyota FJ Cruiser. Readers respond to the blog by posting their own observations, insights, and stories. For many firms, these virtual communities provide various types of information for users, with opportunities for increased interaction via forums, chat rooms, and message boards (Mangold & Faulds, 2009). Such communities provide excellent consumer engagement.

Careful Planning and Little Intervention

As with any communication exercise, it is necessary to plan social media marketing campaigns carefully prior to their launch. Once the process is set in motion, intervention may be disastrous (Berthon et al., 2012;

Kaplan & Haenlein, 2011). Specifically, a truly compelling social media marketing campaign should be able to run independently and develop its own dynamics. Firms should, therefore, develop a comprehensive and robust social media strategy which can generate eWOM dynamically. A good example is Evian, whose Roller-Skating Babies campaign was crowned by the Guinness Book of Records as 'the most viewed advertising spot', with more than 45 million online views. Evian engaged in careful planning before launching the video: it chose the right topic (10 years earlier the company had an advertising campaign based on babies), the right music (a remix of a 30-year-old rap song), and the right messengers. Once the eWOM was unleashed, the company restricted itself to reacting to the eWOM phenomenon, instead of proactively swaying the message.

Create Separate Strategies for Each Channel

Users interact with content in different ways on different social channels (Entrepreneur, 2015). While some users may find pleasure in multiple channels, some have found a single community they love and are sticking with it, even when the lures of a new social channel arise. Some may favour one network today and in six months discover that another meets their needs better. A firm must follow such customers and prospects so it can network in all the right places. It needs to adapt its strategy accordingly while maintaining the same voice across channels. A firm can also seek out inspiration by checking out its top competitors. Investigating what the competitors are doing and how they engage with their followers, can help the firm to meet its consumers' needs better (Dixon, 2000).

Design Products that Create Buzz

To stimulate conversations, products and services should be designed with consumers' desired self-images in mind (Mangold & Faulds, 2009), and must be capable of generating buzz for the business. The key is products and services that are easy to use, fun, intriguing, highly visible, and engaging are more likely to inspire eWOM (Dobele, Lindgreen, Beverland, Vanhamme, & van Wijk, 2007; Dobele, Toleman, & Beverland, 2005). The iPhone is likely to be bought and discussed among consumers who want to be perceived as 'cool'. BMW cars are likely to be discussed and purchased among auto-enthusiasts who value 'the ultimate driving experience'.

Make the Message Memorable and Interesting

The message must be sufficiently memorable and interesting to have the potential to spur eWOM. To make a message more memorable and interesting, Kaplan and Haenlein (2011, p. 257) suggested minor adjustments to the message: 'One option is to rely on true stories about real people ("My brother has a friend, John Doe ..."), which are often more persuasive than corporate advertising. Another option is to use rumours, especially positive ones that reflect well on the person telling them, as they have a particularly high chance of being transmitted to others. Moreover, there are the obvious safe bets like practical short lists (e.g., "The ten best ways to lose weight"), humorous or even hilarious messages'. In general, the message must trigger the receiver's emotions (Dobele et al., 2007), often with an element of surprise, joy, or fear. Also, marketing managers must exercise caution in using messages that may be too provocative or inappropriate.

Leverage Emotional Connections

People discuss topics or subject matters in which they are emotionally connected (Sashi, 2012). Customers may be emotionally connected to finding cures for diseases such as diabetes, cancer, and heart disease. Other emotional linkages include animal rights, environmental issues, child welfare, and education. To leverage emotional connections and generate eWOM, Mangold and Faulds (2009) suggest that organisations can embrace one or more causes important to their consumers. Pepsi, for example, decided not to run any ads during the 2010 Super Bowl. Rather, it spent an estimated US$20 million of the saved budget on the Pepsi Refresh Project, an interactive online platform where users can submit and vote on ideas that have positive impacts and can help make the world a better place (Kaplan & Haenlein, 2011). Ice cream producer Ben and Jerry's supports causes that are important to the founders and its customers. These causes range from investing in the nation's children, supporting family farms, avoiding the consumption of meat from cloned animals, and promoting social and environmental concerns.

Offer Exclusivity

Everyone likes to feel special (Leung, 2013; Mangold & Faulds, 2009). By offering products, information, and special deals exclusively to the consumers, feelings of specialness can be created (Mangold & Faulds, 2009).

For example, Starbucks' stand-out feature on social media is its active online consumer engagement. Starbucks provides its customers with the option to manage and reload their coffee cards online. It also makes it possible for its consumers to locate nearby locations, check out international Starbucks spots, and apply for jobs at its stores. Likewise, Unilever gives its reader's coupons and product samples from Dove's 'Campaign for Real Beauty'. Another example is XBox's Twitter page an interesting resource, with news update, slick advertising photos, and promotions. Gamers can get instantaneous responses to their issues at its @XboxSupport handle. Also, Xbox provides phone numbers and addresses common topics online.

Harness the Power of Storytelling

Stories can be memorable; they provide the connection between facts to make them memorable (Kosara & Mackinlay, 2013). The more memorable the stories are, the more likely they are to generate eWOM (Mangold & Faulds, 2009). Refined stories that depict a real, but still discreetly created image of a business are key to humanising a brand to existing customers, and potential sales leads. For example, the Adidas campaign 'Impossible is nothing' was successful because it communicated a true story; so authentic, that it touched consumers. It left the story of Messi, the concept of the campaign, and the brand impregnated in consumers' memory for years. The phrase 'Impossible is nothing' resonated like a scream of glory in people's ears, and since then Adidas has been associated with that spirit. Such stories create vivid memories that are likely to generate eWOM, as well as traditional WOM (Mangold & Faulds, 2009). Firms must, therefore, harness the power of social media, not just to 'share' content, but to 'tell it', and to 'narrate' the brand. Social media platforms need to be seen as a conversation channel; a transmission of experiences, values, dreams, feelings, failures, successes, and truths.

Solicit Feedback

Feedbacks are important for consumer engagement. Consumers become more involved with products and organisations when they can submit feedback (Mangold & Faulds, 2009). For example, Toyota solicits and gets feedbacks from consumers in the form of accolades, criticism, and helpful suggestions on a broad range of issues through its 'Open Road

Blog'. The feedback provides a sense of community in which open communications are stimulated, and customer engagement is boosted. An organisation can leverage its network, for example, before launching products to solicit the feedback of its virtual communities (Heller Baird & Parasnis, 2011). One of the greatest benefits of social media is that a firm's network probably consists of not only existing customers, but also target customers, prospects, and industry influencers. Indeed, a firm can get a wide range of feedback on new products (i.e., whether they are worth launching in the first place; whether they are easy to use; or what should be next in the development queue) and get the all-important community buy-in for greater adoption rates at launch time.

In line with the theory of social influence (Friedkin, 2006), people are strongly influenced by how they perceive their relationship with the influencer. Likewise, consumers are more likely to communicate when they have a relationship with the product or service. In particular, some companies have been encouraging consumers to submit photos or recorded footage of their product in action. For example, a promotion for Procter and Gamble's quick-clean product, *Swiffer* challenged consumers to make short videos recounting 'how they left their old cleaning method for a new romance with Swiffer'. Then, the top ten videos were posted on YouTube and consumers were asked to vote on the winner. The contest sparked a huge amount of attention. Such online voting provides Internet users with a sense of ownership and increased engagement.

Intensify Social Media Presence to Increase Reach

Search engines are beginning to adopt signals from social media sites into their search ranking design (e.g., Bing Social Search, TripAdvisor). Social media presence can therefore greatly increase a firm's search rankings (Demers, 2015). Recent studies (i.e., Ghose, Ipeirotis, & Li, 2014) have found that a consumer utility-based ranking mechanism on product search engines, that integrates multidimensional consumer preferences and signals from social media sites leads to significant surplus gain for consumers. As with most search ranking strategies, there is one basic principle: the better experience you give your users, the higher you rank in Google. Going by the norm of reciprocity (Burger, Sanchez, Imberi, & Grande, 2009), if a firm gives its users a better experience, it has opened the widest possible channel for new potential customers to find its brand.

Include the Rest of the Marketing Communications Mix

Despite all the advantages of social media in marketing communications, it needs to be accompanied by concerted efforts with the rest of the marketing communications mix (Mangold & Faulds, 2009). The buzz surrounding social media marketing usually dies out quickly. 'Nothing tends to be more boring than yesterday's news', the old saying goes. To maintain the eWOM momentum, social media marketing must be complemented with more traditional forms of communication. Kaplan and Haenlein (2011, p. 261) illustrated this by relating the story of Wilkinson's Fight for Kisses advertisement as follows: 'To support the launch of its newly-developed disposable razor, Quattro Titanium, the company relied on a viral marketing story about a baby fighting his father for kisses from the baby's mother. This campaign consisted of an animated video and an interactive computer game, combined with a series of press announcements, radio spots, and sponsorship of the France—Ireland rugby match that took place during the same time period. Despite its limited budget of only s62,000 (about US$90,000), the campaign was a huge success and resulted in a five-percentage-point market share increase within the target group'. Social media marketing promotes satisfactory results when utilised in combination with the rest of the marketing communications mix.

The literature has also demonstrated that social media play a significant role in driving traditional media activity (Kaplan & Haenlein, 2011; Mangold & Faulds, 2009). For example, Stephen and Galak (2012) analysed 14 months of daily sales and media activity data from a microlending marketplace using a multivariate autoregressive model. The authors found that both traditional and social earned media affect sales. However, because of the higher frequency of social earned media activity after adjusting for event frequency, social earned media's sales elasticity is significantly higher than traditional earned media's. Further, social earned media tends to play a significant role in driving traditional earned media activity. Organisations should, therefore, combine social media and the traditional media in order to achieve marketing success.

Managerial Implications

The study has shown that, in social media marketing, emphasis needs to be placed on creating and adopting effective strategies that can generate eWOM and make the brand go viral. The strategies synthesised in this study will, therefore, help managers to modify their marketing input. The budding field of social media marketing tends to borrow many of the traditional off-line techniques for managing interpersonal influence (i.e., stimulating and creating opinion leaders) while at the same time creating new techniques aided by the special characteristics of social media. Marketing managers must understand that their consumers are going online in increasing numbers and that, on the social media, they are being exposed to and influenced by their competitors. Marketing managers should, therefore, take the lead in understanding and utilising the emerging technology and strategies of social media marketing.

This study has identified a series of available strategies for harnessing the power of the social media for firms to stimulate their marketing efforts. Firstly, as with any communication exercise, it is necessary to plan social media marketing campaigns extensively before their launch. Once the eWOM is launched, the company should limit itself to reacting to the eWOM phenomenon, instead of proactively swaying the message. Additionally, in contrast to traditional marketing, which involves limited customer participation, managers need to develop new forms of advertorial content that encourage user engagement. Marketing managers are more likely to generate eWOM if they continue to inspire their audience; make their content shareable, fun and useful; use great graphics and video; and cross-promote to other channels. As well, it is necessary to create separate strategies for each channel because users interact with content differently on every social channel. A firm must follow such customers and prospects so it can network in all the right places.

Creating successful social media marketing requires the right people to spread the message. To effectively transform an ordinary message into an eWOM phenomenon, therefore, a firm must employ the three identified groups of messengers (i.e., market mavens, social hubs, and salespeople) to spread the message. While spreading the message, organisations should create virtual communities of individuals with common interests, centred on shared interests and values. Such communities provide excellent consumer engagement.

To further stimulate eWOM, products and services should be designed with consumers' desired self-images in mind to create buzz. The message must be sufficiently memorable and interesting to have the potential to spur eWOM. Marketing messages must trigger the consumer's emotions. Firms must leverage emotional connections because people discuss things to which they are emotionally connected. Likewise, people like to feel special. To leverage on this, managers must provide exclusivity.

Also, marketing managers must harness the power of stories. Stories can be memorable. The more memorable the stories are, the more likely they are to generate eWOM. Feedbacks are, as well, important for consumer engagement. Indeed, a firm can get a wide range of feedback on new and existing products and get the all-important community buy-in for all the ranges of its products. Firms must also increase social media presence in order to greatly increase their search rankings. If a firm gives its users a better experience, it has opened the widest possible channel for new potential customers to find its brand. Also, social media marketing must be complemented with the rest of the marketing communications mix. To maintain the eWOM momentum, therefore, firms should combine social media and the traditional media in their marketing campaigns.

Limitations and Implications for Future Research

Given the conceptual nature of the study and the fact that social media is an emerging field of research, the strategies should be considered exploratory. Although many global companies have successfully created eWOM with the identified strategies, surveys should be carried out to provide a better-grounded view of the social media strategies. Also, a longitudinal examination of social media strategies is necessary in order to determine how firms adjust their strategies over time.

Further, the strategies offered in this chapter suggest a number of interesting research questions. What strategies are most consistent with successful social media marketing? Some strategies have been offered, but that list is by no means exhaustive. What range of marketing strategies and associated practices flows from social media? How do the connective processes in social media contribute to the development of marketing practices?

Some social media strategies emerge over time and are not always preplanned. Investigation into those strategies means focusing qualitatively on the connective capabilities of social media and on the psychology of e-conversations that can lead to transformative change in marketing. One way to approach some of these questions would be to construct some high-level, exploratory hypotheses tying a marketing approach on social media to perceived strategic resources to organisational outcomes. One could construct a hypothesis concerning the significance of a social media strategy, and another pertaining to the degree of usage of the strategy, covering the intention and execution. For instance:

- Competitive advantage flows from appropriate marketing strategies on social media.
- Greater usage of social media marketing strategies is associated with greater relative industry performance.

A possible method for exploring these hypotheses might be to pick comparator companies, who have relatively equal strategic positions in the same industry, and gather qualitative data on their respective social media marketing schemes. Qualitative and quantitative comparisons of high-and low-performing firms in the same industry could as well be constructed to explore the hypothesis and to help in the construction of others more specific. Most significant and perhaps generalisable findings out of such a study would be not just the particular strategies themselves but the mechanism and rationale within the firm of drafting such strategies (e.g., the participants, time frames, conversations, and inputs).

Beyond the above macro-oriented questions, managers should, on a practical level, initiate their own studies focused on their specific products and services to assess the level of current eWOM and explore different strategies to increase it.

References

Agnihotri, R., Dingus, R., Hu, M. Y., & Krush, M. T. (2016). Social media: Influencing customer satisfaction in B2B sales. *Industrial Marketing Management, 53*, 172–180.

Akehurst, G. (2009). User generated content: The use of blogs for tourism organisations and tourism consumers. *Service Business, 3*(1), 51.

Bawden, D., & Robinson, L. (2009). The dark side of information: Overload, anxiety and other paradoxes and pathologies. *Journal of Information Science, 35*(2), 180–191.

Berinato, S. (2010). Six ways to find value in Twitter's noise. *Harvard Business Review, 88*(6), 34–35.

Berthon, P. R., Pitt, L. F., Plangger, K., & Shapiro, D. (2012). Marketing meets Web 2.0, social media, and creative consumers: Implications for international marketing strategy. *Business Horizons, 55*(3), 261–271.

Bowen, G., & Bowen, D. (2016). Social media: Strategic decision making tool. In *Competitive social media marketing strategies* (pp. 94–111). IGI Global.

Bruce, N. I., Foutz, N. Z., & Kolsarici, C. (2012). Dynamic effectiveness of advertising and word-of-mouth in the sequential distribution of short life cycle products. *Journal of Marketing Research, 49*(4), 469–486.

Buhalis, D., & O'Connor, P. (2005). Information communication technology revolutionizing tourism. *Tourism Recreation Research, 30*(3), 7–16.

Burger, J. M., Sanchez, J., Imberi, J. E., & Grande, L. R. (2009). The norm of reciprocity as an internalized social norm: Returning favors even when no one finds out. *Social Influence, 4*(1), 11–17.

Chauhan, K., & Pillai, A. (2013). Role of content strategy in social media brand communities: A case of higher education institutes in India. *Journal of Product & Brand Management, 22*(1), 40–51.

Chevalier, J. A., & Mayzlin, D. (2006). The effect of word of mouth on sales: Online book reviews. *Journal of Marketing Research, 43*(3), 345–354.

Chu, S. C., & Kim, Y. (2011). Determinants of consumer engagement in electronic word-of-mouth (eWOM) in social networking sites. *International Journal of Advertising, 30*(1), 47–75.

Chung, C., & Austria, K. (2010). Social media gratification and attitude toward social media marketing messages: A study of the effect of social media marketing messages on online shopping value. *Proceedings of the Northeast Business & Economics Association.* (pp. 581–586).

Constantinides, E. (2009). Social Media/Web 2.0 as marketing parameter: An introduction [Online]. *International Marketing Trends Conference*, Paris 15–17 January 2009. Retrieved November 26, 2009, from http://marketing-trends-congress.com/2009_cp/Materiali/Paper/Fr/Constantinides.pdf.

Crotts, J. (1999). Consumer decision making and prepurchase information search. In *Consumer behavior in travel and tourism* (pp. 149–168). New York: The Haworth Hospitality Press.

Darley, W. K., Blankson, C., & Luethge, D. J. (2010). Toward an integrated framework for online consumer behavior and decision making process: A review. *Psychology & Marketing, 27*(2), 94–116.

Demers, J. (2015). 6 social media practices that boost SEO. Retrieved January 22, 2017, from http://www.forbes.com/sites/jaysondemers/2015/01/27/6-social-media-practices-that-boost-seo/2/#d131b5564472.

Dholakia, U. M., & Durham, E. (2010). One café chain's Facebook experiment. *Harvard Business Review, 88*(3), 26.

Dixon, N. M. (2000). *Common knowledge: How companies thrive by sharing what they know*. Harvard Business School Press.

Dobele, A., Lindgreen, A., Beverland, M., Vanhamme, J., & van Wijk, R. (2007). Why pass on viral messages? Because they connect emotionally. *Business Horizons, 50*(4), 291–304.

Dobele, A., Toleman, D., & Beverland, M. (2005). Controlled infection! Spreading the brand message through viral marketing. *Business Horizons, 48*(2), 143–149.

Dolan, R. (2016). Social media: Are Facebook fans really 'engaging' with our wine brands?: A case study of Australian wine brand Facebook pages. *Wine & Viticulture Journal, 31*(1), 67.

Effing, R., & Spil, T. A. (2016). The social strategy cone: Towards a framework for evaluating social media strategies. *International Journal of Information Management, 36*(1), 1–8.

Engel, J. F., Kegerreis, R. J., & Blackwell, R. D. (1969). Word-of-mouth communication by the innovator. *Journal of Marketing, 33*(3), 15–19.

Entrepreneur.Com. (2015, May 1). Adopting a winning social strategy. Retrieved from www.entrepreneur.com

Erdoğmuş, İ. E., & Cicek, M. (2012). The impact of social media marketing on brand loyalty. *Procedia-Social and Behavioral Sciences, 58*, 1353–1360.

Feick, L. F., & Price, L. L. (1987). The market maven: A diffuser of marketplace information. *Journal of Marketing, 51*(1), 83–97.

Felix, R., Rauschnabel, P. A., & Hinsch, C. (2017). Elements of strategic social media marketing: A holistic framework. *Journal of Business Research, 70*, 118–126.

Fotis, J. N. (2015). *The Use of social media and its impacts on consumer behaviour: The context of holiday travel.* Doctoral dissertation, Bournemouth University.

Foux, G. (2006, May 8). Consumer-generated media: Get your customers involved. *Brand Strategy*, pp. 38–39.

Friedkin, N. E. (2006). *A structural theory of social influence* (Vol. 13). Cambridge University Press.

Gallaugher, J., & Ransbotham, S. (2010). Social media and customer dialog management at Starbucks. *MIS Quarterly Executive, 9*(4), 197–212.

Ghose, A., Ipeirotis, P. G., & Li, B. (2014). Examining the impact of ranking on consumer behavior and search engine revenue. *Management Science, 60*(7), 1632–1654.

Godes, D., & Mayzlin, D. (2004). Using online conversations to study word-of-mouth communication. *Marketing Science, 23*(4), 545–560.

Goldenberg, J., Libai, B., & Muller, E. (2001). Talk of the network: A complex systems look at the underlying process of word-of-mouth. *Marketing Letters, 12*(3), 211–223.

Hackworth, B. A., & Kunz, M. B. (2011). Health care and social media: Building relationships via social networks. *Academy of Health Care Management Journal, 7*(2), 1.

Heller Baird, C., & Parasnis, G. (2011). From social media to social customer relationship management. *Strategy & Leadership, 39*(5), 30–37.

Hennig-Thurau, T., Gwinner, K. P., Walsh, G., & Gremler, D. D. (2004). Electronic word-of-mouth via consumer-opinion platforms: What motivates consumers to articulate themselves on the internet? *Journal of Interactive Marketing, 18*(1), 38–52.

Hudson, S., & Thal, K. (2013). The impact of social media on the consumer decision process: Implications for tourism marketing. *Journal of Travel & Tourism Marketing, 30*(1–2), 156–160.

Internet Live Stats. (2019, July 18). Internet usage and social media statistics. Retrieved July 18, 2019, from https://www.internetlivestats.com.

Kaplan, A. M., & Haenlein, M. (2011). Two hearts in three-quarter time: How to waltz the social media/viral marketing dance. *Business Horizons, 54*(3), 253–263.

Katz, E., & Lazarsfeld, P. F. (1955). *Personal influence: The part played by people in the flow of mass communications*. Glencoe, IL: The Free Press.

Kaushik, R. (2012, March). Impact of social media on marketing. *IJCEM International Journal of Computational Engineering & Management, 15*(2), 2230–7893.

Kietzmann, J. H., Hermkens, K., McCarthy, I. P., & Silvestre, B. S. (2011). Social media? Get serious! Understanding the functional building blocks of social media. *Business Horizons, 54*, 241–251.

Kirtiş, A. K., & Karahan, F. (2011). To be or not to be in social media arena as the most cost-efficient marketing strategy after the global recession. *Procedia-Social and Behavioral Sciences, 24*, 260–268.

Kosara, R., & Mackinlay, J. (2013). Storytelling: The next step for visualization. *Computer, 46*(5), 44–50.

Kumar, V., Choi, J. B., & Greene, M. (2016). Synergistic effects of social media and traditional marketing on brand sales: Capturing the time-varying effects. *Journal of the Academy of Marketing Science, 45*(2), 1–21.

Leung, L. (2013). Generational differences in content generation in social media: The roles of the gratifications sought and of narcissism. *Computers in Human Behavior, 29*(3), 997–1006.

Lipsman, A., Mudd, G., Rich, M., & Bruich, S. (2012). The power of 'like'. How brands reach (and influence) fans through social-media marketing. *Journal of Advertising Research, 52*(1), 40–52.

Mangold, W. G., & Faulds, D. J. (2009). Social media: The new hybrid element of the promotion mix. *Business Horizons, 52*(4), 357–365.

Moscovici, S. (1980). Toward a theory of conversion behavior. *Advances in Experimental Social Psychology, 13*, 209–239.

Neti, S. (2011). Social media and its role in marketing. *International Journal of Enterprise Computing and Business Systems, 1*(2), 1–15.

Onishi, H., & Manchanda, P. (2012). Marketing activity, blogging and sales. *International Journal of Research in Marketing, 29*(3), 221–234.

Pan, B., & Crotts, J. C. (2012). Theoretical models of social media, marketing implications, and future research directions. In *Social media in travel, tourism and hospitality: Theory, practice and cases* (pp. 73–85). Surrey, UK: Ashgate.

Pfeffer, J., Zorbach, T., & Carley, K. M. (2014). Understanding online firestorms: Negative word-of-mouth dynamics in social media networks. *Journal of Marketing Communications, 20*(1–2), 117–128.

Rodríguez-Ardura, I., & Meseguer-Artola, A. (2010). Toward a longitudinal model of e-commerce: Environmental, technological, and organizational drivers of B2C adoption. *The Information Society, 26*(3), 209–227.

Rolls, K., Hansen, M., Jackson, D., & Elliott, D. (2016). How health care professionals use social media to create virtual communities: An integrative review. *Journal of Medical Internet Research, 18*(6), e166.

Sashi, C. M. (2012). Customer engagement, buyer-seller relationships, and social media. *Management Decision, 50*(2), 253–272.

Selina, D., & Milz, T. (2009). Social media will be a driving force for relationship development. *Credit Union Journal, 13*(32), 16.

Sharma, A., & Sheth, J. N. (2004). Web-based marketing: The coming revolution in marketing thought and strategy. *Journal of Business Research, 57*(7), 696–702.

Stephen, A. T., & Galak, J. (2012). The effects of traditional and social earned media on sales: A study of a microlending marketplace. *Journal of Marketing Research, 49*(5), 624–639.

Thorson, K. S., & Rodgers, S. (2006). Relationships between blogs as eWOM and interactivity, perceived interactivity, and parasocial interaction. *Journal of Interactive Advertising, 6*(2), 5–44.

Trusov, M., Bucklin, R., & Pauwels, K. (2009). Effects of word-of-mouth versus traditional marketing: Findings from an internet social networking site. *Journal of Marketing, 73*(5), 90–102.

Tsimonis, G., & Dimitriadis, S. (2014). Brand strategies in social media. *Marketing Intelligence & Planning, 32*(3), 328–344.

Tuten, T. L., & Solomon, M. R. (2014). *Social media marketing*. Sage.

Urban, G. (2003). Customer advocacy: Is it for you? *Center for Ebusiness at MIT* [Online]. Retrieved March 25, 2011, from http://ebusiness.mit.edu/research/papers/175_Urban_Trust.pdf

Van den Bulte, C., & Lilien, G. L. (2001). Medical innovation revisited: Social contagion versus marketing effort. *American Journal of Sociology, 106*(5), 1409–1435.

van Raaij, W. F. (1998). Interactive communication: Consumer power and initiative. *Journal of Marketing Communications, 4*(1), 1–8.

Villanueva, J., Yoo, S., & Hanssens, D. M. (2008). The impact of marketing induced versus word-of-mouth customer acquisition on customer equity growth. *Journal of Marketing Research, 45*(1), 48–59.

Vollmer, C., & Precourt, G. (2008). *Always on: Advertising, marketing, and media in an era of consumer control*. New York: McGraw-Hill.

Weber, L. (2009). *Marketing to the social web: How digital customer communities build your business*. John Wiley & Sons.

CHAPTER 5

Social Media Technologies and Export Marketing

Frederick Avornyo, Mahmoud Abdulai Mahmoud, Ogechi Adeola, Robert Ebo Hinson, and Richard Boateng

Introduction

Social media marketing is a major enabler of small- and medium-scale enterprises (SMEs) venturing into the international markets (Berthon, Pitt, Plangger, & Shapiro, 2012; Okazaki & Taylor, 2013). Social media, as a resource to drive marketing, enables small firms with limited resources to improve their international communication, be more efficient with market transactions with customers abroad, achieve foreign customer satisfaction and loyalty, and generally develop international network relationships (Alarcón-del-Amo, Rialp, & Rialp, 2015). These qualities, defined as international social media competencies by Alarcón-del-Amo et al. (2015), lead to the rapid internationalisation of SMEs through exports

F. Avornyo (✉) • M. A. Mahmoud • R. E. Hinson • R. Boateng
University of Ghana Business School, Accra, Ghana
e-mail: mamahmoud@ug.edu.gh; richboateng@ug.edu.gh

O. Adeola
Pan-Atlantic University, Lagos, Nigeria
e-mail: oadeola@lbs.edu.ng

B. George, J. Paul (eds.), *Digital Transformation in Business and Society*, https://doi.org/10.1007/978-3-030-08277-2_5

(Alarcón-del-Amo, Rialp, & Rialp, 2017). SMEs generally struggle to internationalise due to inadequate resources (Paul, Parthasarathy, & Gupta, 2017). A key challenge is the ability to communicate effectively with foreign customers (Paul et al., 2017), but this appears to be changing with the advent of social media, justifying the position by scholars that social media is changing the conduct of business, and internet technology, in general, has reduced the cost of international business (Galati, Crescimanno, Tinervian, & Fagnani, 2017; Kaplan & Haenlein, 2010; Paniagua, Korzynski, & Mas-Tur, 2016; Sinkovics, Sinkovics, & Jean, 2013). The changes in the conduct of business triggered by social media, therefore, require that more scholarly attention should be directed towards the phenomenon.

The ability of the internet technology to facilitate export has been attested to and gained much scholarly attention (Gregory, Ngo, & Karavdic, 2017; Hinson, 2010; Sinkovics et al., 2013). However, high cost and reliability of internet technology (emails and websites) have been found to be impairing its successful adoption by SMEs in developing countries (Boateng, Heeks, Molla, & Hinson, 2011; Hinson, 2010). This narrative could also be changing with the emergence of social media, the revolutionary internet-based technology, which is offering firms new ways of interacting with their stakeholders (Alarcón-del-Amo et al., 2015; Kaplan & Haenlein, 2010). Importantly, social media is cheaper and easier to use, provides global reach, breaks time and distance barriers, makes communication immediate and engenders networking between the firm and the foreign consumer (Alarcón-del-Amo, Rialp, & Rialp, 2016; Berthon et al., 2012; Genc & Oksuz, 2015; Odoom, Anning-Dorson, & Acheampong, 2017; Okazaki & Taylor, 2013). The ability to access social media via the mobile phone, which has high penetration rate, is considered an added advantage, making it an important and easy-to-use resource (Packer, 2011) for owner-managers of SMEs.

The impact of social media resource on exports is indisputable (Alarcón-del-Amo et al., 2017), as affirmed by recent findings by a few scholars. The global logistics company UPS also confirms that 'social media is having a profound effect on exporters, and in particular, on small businesses, that often find their goods in demand from overseas customers long before they have thought about breaking into international markets' (compass.ups.com, n.d.). This claim by UPS is consistent with international marketing literature which suggests that social media is enabling otherwise cash-strapped firms or SMEs, which could not afford physical presence on the international market, to be in a position to have their products on the international market (Alarcón-del-Amo et al., 2017).

However, some authors argue that firms' mere presence on social media cannot be the reason for the rapid internationalisation of SMEs (Alarcón-del-Amo et al., 2017), rather success is achieved when firms are able to align the social media resource and other complementary resources, to the needs of consumers in foreign markets (Seggie & Griffith, 2008). This assertion is further buttressed by the argument that technology on its own cannot lead to performance, while firm resources alone can neither produce significant gains in firm performance (Bharadwaj, 2000; Trainor, Andzulis, Rapp, & Agnihotri, 2014). Scholars posit that firm performance is possible when technological resources such as social media is combined with firm's strategic resources to achieve distinctive capabilities (Alarcón-del-Amo et al., 2017; Trainor et al., 2014). This is consistent with the assertion by the Resource-Advantage (R-A) theorists, who argue that firms combine resources, both tangible and intangible, to achieve superior performance (Hunt & Arnett, 2003; Hunt, & Morgan, 1995).

Scholars have empirically verified that social media utilisation leads to firm performance (Luo, Zhang, & Duan, 2013; Paniagua & Sapena, 2014) and export performance specifically (Alarcón-del-Amo et al., 2017). However, the issue as to what resource combination SMEs on social media are aligning to the needs of foreign consumers, to achieve this export performance, appears not to have been adequately addressed by extant literature (Alarcón-del-Amo et al., 2017). Social media is considered nascent (Killian & McManus, 2015) and social media application to international marketing is still considered emerging (Okazaki & Taylor, 2013), and therefore not surprising that research into social media marketing and export is scanty (Paniagua et al., 2016). This chapter, therefore, seeks to understand which resources firms are integrating with social media to achieve superior performance. The study is linked to the central focus of the resource-advantage theory (R-A) theory and is, therefore, premised on the approach. The rest of the chapter looks at the theoretical foundation of the study, a review of the literature on social media marketing, a discussion on social media export marketing capabilities, social media and export performance measurement, proposes a framework and presents its conclusions.

Theoretical Foundation

Social media is a resource (Paniagua et al., 2016; Wang & Kim 2017) which firms efficiently and effectively combine with other resources to achieve superior performance (Alarcón-del-Amo et al., 2016; Wang & Kim, 2017). This assertion is in sync with the R-A theory that firms are combiners of dynamic and heterogeneous resources to create a comparative advantage leading to competitive advantage and superior financial performance (Hu & Wang, 2009; Hunt, 2011; Hunt & Morgan, 1995; Thoeni, Marshall, & Campbell, 2016). The theory places emphasis on innovation, market segments, heterogeneous firm resources, a comparative market advantage/disadvantage in resources and a marketplace position of competitive advantage/disadvantage (Hunt, 2001, 2012). R-A theory enumerates seven resources which it argues are available to firms. They include financial, physical, legal, human, organisational, informational and relational (Barua & Michie, 2014; Hunt 2001, 2012). The theory posits that firms combine both tangible and intangible resources to deliver value for a segment of the market which is akin to what exporting SMEs present on social media do. They combine social media marketing with other firm resources to deliver value to customers in the international segment of the market.

R-A theory was propounded in 1995 by Shelby Hunt and Robert Morgan in an article titled 'The Comparative Advantage Theory of Competition'. It is an interdisciplinary and evolutionary general theory of competition that evolved from ten different research traditions (Seggie & Griffith, 2008) and has 12 foundational premises and a structure (Hunt & Arnett, 2003). Hunt and Arnett posit that R-A is

> an evolutionary, disequilibrium-provoking, process theory of competition, in which innovation and organisational learning are endogenous, firms and consumers have imperfect information, and in which entrepreneurship, institutions, and public policy affect economic performance.

This suggests that firms will avail themselves of innovations (such as social media), and as they combine resources available to them to achieve superior performance (such as in export), they will learn internally (since organisational learning is endogenous) which resources would deliver better outcomes, so they can determine, in which resources more investment commitment should be made (Rest & Roper, 2013).

The theory also argues that firms combine an assortment of resources available to them in order to have a comparative advantage in resources

(Hunt 2012; Hunt and Morgan, 1995, 1997). For firms, this will result in a marketplace position of competitive advantage and hence superior performance (Hunt, 2012; Hunt & Arnett, 2003). On the other hand, a comparative disadvantage in resources will lead to a marketplace position of competitive disadvantage and inferior performance, with a third position being where there is parity (Hunt, 2011). It, therefore, becomes relevant for firms to appreciate what resource combinations deliver comparative advantage and ultimately superior performance and which ones deliver the opposite outcomes.

Since its seminal paper in 1995, several other researchers have contributed to the theory by using it as basis to explain various phenomena including resource matching and globalisation of services (Seggie & Griffith, 2008), to demonstrate that Corporate Social Responsibility is core to firm operations (Hu & Wang, 2009), while others have extended the theory into firm's pricing strategy (van der Rest & Roper, 2013). However, a review of the extant literature on R-A theory did not reveal its usage to explain the social media marketing and export performance phenomenon. This is in spite of the fact that R-A theory is considered a marketing strategy theory, while social media marketing has become an integral part of marketing strategy for firms (Akar & Topçu, 2011; Barua & Michie, 2014; Schlegelmilch, 2011). Additionally, the list of seven resources as provided for by the R-A theory cannot be exhaustive as marketing practice is dynamic (Grönroos, 1994). The list of resources could, therefore, be extended. The chapter also argues that not all the seven resources can be of equal importance to a firm. The Competitive Position Matrix of R-A theory demonstrates that some resources will lead to marketplace position of competitive advantage, while others may not (Hunt, 2001). On that basis, it is argued that some resources may be of higher importance than others. The most relevant resources can be classified as core and the rest secondary. This qualification of the resources is considered relevant, as the complementarity or otherwise of the resources could influence how they are converted into capabilities and eventually to superior performance (Trainor et al., 2014).

Literature Review

Social media marketing is one of the contemporary marketing concepts, and business practitioners are seeking to understand how to apply it effectively in order to achieve superior performance (Packer, 2011). It has unique features which could explain the special traction it has received in recent times in preference to the older internet-based technologies (Elena, 2016; Mangold & Faulds, 2009). These features include the easy and quick unfettered access to the global market, content co-creation, real-time interactivity and engagement between the firm and the consumer as well as among consumers (Martin, Goncalves, Oliveira, Cota, & Branco, 2016; Wang & Kim, 2017). Social media marketing emerged following the advent of social media platforms such as Facebook, Twitter, Instagram and several other forms such as Pinterest, YouTube, WhatsApp, LinkedIn, Google+, Snapchat (Paniagua et al., 2016; Socialpilot.co, n.d.). It consists of text, pictures, videos and networks (Berthon et al., 2012). Attempts have been made to categorise the various social media platforms into

1. collaborative projects such as Wikipedia,
2. blogs and user-generated content communities such as YouTube,
3. social networking sites such as Facebook,
4. microblogs such as Twitter,
5. virtual game worlds such as EverQuest and
6. virtual social worlds such as Second Life (Berthon et al., 2012; Kaplan & Haenlein, 2010; Okazaki & Taylor, 2013).

Social media is a Web 2.0 application which allows for two-way and many-to-many ways communication (Ananda, Hernández-García, & Lamberti, 2016). Scholars have explained Web 2.0 to be the interactive component of Web 1.0 (Yadav & Rahman, 2017), as Web 1.0 is only a one-way application such as websites (Constantinides, 2014). One of the scholars cited as a pioneer, Tim O'Reilly (Berthon et al., 2012) sees social media in the light of Web 2.0 and defines it in a technical sense as 'a set of principles and practices that tie together a veritable solar system of sites that demonstrate some or all of those principles, at a varying distance from that core' (O'Reilly, 2007). Subsequently, Kaplan and Haenlein (2010) simplify it as an internet-based application that allows for the creation and exchange of user-generated content, but they also draw the linkage with Web 2.0.

Though the definitions found in extant literature are varied, there are commonalities which include the fact that social media is an internet-based technology (Kaplan & Haenlein, 2010), modelled on Web 2.0 (Constantinides, 2014; Wang & Kim, 2017) that allows users to interact, share information and co-create content (Berthon et al. 2012). Based on these various definitions, and others in extant literature, this chapter conceptualises social media as a social Web 2.0 platform that allows for real-time interactivity and content co-creation among users, in the form of text, picture, video, audio and network (Berthon et al., 2012; Constantinides, 2014; Kaplan & Haenlein, 2010).

Social media marketing on the other hand involves the use of social media technology for firms' promotional activities (Akar & Topçu, 2011; Kaplan & Haenlein, 2010) to achieve a firm's goal by creating value for its various stakeholders (Felix, Rauschnabel, & Hinsch, 2017; Tuten & Solomon, 2015) and for relationship building (Chang et al., 2015). Based on these cited expert views (i.e. Chang et al., 2015; Felix et al., 2017), we define social media marketing as the use of social media for marketing activities, including research, promotion, distribution and engagement with and among stakeholders, for value creation.

Thematic Areas

We reviewed 133 articles to address various issues which were synthesised into ten thematic areas for the purpose of this chapter. A significant number of articles (27.8%) examined factors that influence social media adoption in order to determine how firms can be successful with its application (Chikandiwa, Contogiannis, & Jembere, 2013; Galati et al., 2017). Some examined social media and branding (Bruhn, Schoenmueller, & Schäfer, 2012; Nguyen, Yu, Melewar, & Chen, 2015), and how social media could predict consumer behaviour (Godey et al., 2016; Kim & Drumwright, 2016). Other scholars delved into how social media influences electronic word of mouth and its impact on the operations of firms (Erkan & Evans, 2016; Teng, Khong, Chong, & Lin, 2017). Other thematic areas identified include social customer relationship management (Malthouse, Haenlein, Skiera, Wege, & Zhang, 2013; Trainor et al., 2014; Wang & Kim, 2017), social media and marketing communications (Mangold & Faulds, 2009; Wang, Pauleen, & Zhang, 2016), and social media marketing strategy (Ananda et al., 2016; Felix et al., 2017). There are also articles that addressed how social media impact firm performance (Luo et al., 2013; Moro, Rita, & Vala, 2016), social media challenges (Kaplan & Haenlein, 2010; Schultz & Peltier, 2013) and finally social media and internationalisation (Alarcón-del-Amo et al., 2017; Berthon et al., 2012; Paniagua et al., 2016).

Three thematic areas namely social media adoption, social media and branding as well as social media and consumer behaviour jointly accounted for 62.4% of the literature examined, while social media marketing strategy, social media challenges and social media and internationalisation, which included export, were among the least researched areas. This finding is consistent with the suggestion by Paniagua et al. (2016) and Alarcón-del-Amo et al. (2017) that the theme, social media and export is under-researched and requires more scientific research attention.

Social Media Adoption and Exporting SMES

Businesses have adopted social media to enhance their marketing operations. Some have achieved success (Felix et al., 2017) others have not witnessed the corresponding improvement in firm performance (Zhu & Chen, 2015). These two outcomes have engendered scholars to research into factors that influence a successful adoption of social media (Martin et al. 2016; Packer, 2011), and the various adoption models and how they impact success (Chikandiwa et al., 2013; Felix et al., 2017).

Chikandiwa et al. (2013) identify two social media adoption models. The first is based on how social media is managed within the organisation; whether it is decentralised or not. They list centralised, distributed, 'coordinated hub and spoke', as well as 'multiple hub and spoke', as forms of adoption frameworks. According to the authors, the 'coordinated hub and spoke' is the adoption model where tactical decision considering a single higher department takes the adoption and usage of social media, but the implementation is decentralised at the various departments within the firm. In the case of 'multiple hub and spoke', again the tactical decision is a top management action, but the implementation is further decentralised beyond departments to smaller units such as product lines. However, in the case of the centralised model, both tactical decision and implementation is the sole responsibility of a dedicated department or unit, and the reverse holds for the distributed model. The second category of social media adoption model by Chikandiwa et al. (2013) examines the various social media adoption stages. They include the social media strategy learning curve, social media adoption curve, social engagement journey and social media maturity model. The authors then explicate the intricacies of each stage and how firms behave at those stages of maturity in the adoption of social media.

Felix et al. (2017) also propose a four-dimensional model which looks at the scope, the culture, the structure and the governance. The scope is concerned with whether firms use social media to engage a few stakeholders or it is an imbibed and holistic philosophy that ensures that both employees and external stakeholders are engaged using social media. In terms of culture, the scholars seek to address the question whether a firm adopts a conservative or modern approach towards the use of social media. Structure, on the other hand, is similar to Chikandiwa et al.'s (2013) first category, which looks at the centralisation or otherwise. Felix et al. (2017) however use the term hierarchies to denote centralisation, and networks to represent a decentralised system of social media usage in a firm. The last dimension is governance. Where there are strict regulations as to who is permitted to use social media within the firm, it is referred to as autocracy, but where there is no strict compliance, it is termed anarchy. Felix et al. (2017) however suggest that in adopting social media, firms should be flexible and have a blend of the dimensions and also use influencers within the firm to lead the social media adoption model.

It should be noted that the studies conducted by Chikandiwa et al. (2013) and Felix et al. (2017) are not specific to SMEs, particularly exporting SMEs. SME-biased researches, however, are extremely important because they dominate most economies (Opoku, Abratt, Bendixen, & Pitt, 2007; Stoian, Rialp, & Rialp, 2011) and more importantly exhibit unique characteristics such as the dominance of owner-managers, inadequate resources to have a full-fledged organisational set up, but quick and flexible in decision making (Lu & Julian, 2008; Mehrtens, Cragg, & Mills, 2001; Paul et al., 2017). This chapter posits that the decision as to which social media adoption model to opt for is constrained by the social media competence of the owner-manager, the strength and quality of the human resource available to the firm, and financial resources to outsource the role of managing the firm's social media marketing implementation. A study conducted on SMEs' social media adoption model, should, therefore, bring out these perspectives unique to SMEs. This chapter, therefore, takes the position that the unique characteristics of SMEs would influence their social media adoption model. A resource-strapped but social media savvy owner-manager of an exporting SME, for instance, may also be constrained by time availability, to continually update its page, respond and regularly engage with online customers who may be of different time zones.

The argument that SMEs could have different social media adoption model is supported by Genc and Oksuz (2015) who clearly demonstrate the large-scale and small-scale business dichotomy by indicating that small businesses have a more liberal approach to social media compared to their large-scale counterparts. Where there is no difference between large and smaller firms is the compelling reason for social media adoption which is the attainment of superior performance (Genc & Oksuz, 2015; Martin et al., 2016), and this can be achieved if firms succeed in effectively and efficiently combining resources available to them. The mere presence of exporting SMEs on any of the social media platforms would not guarantee them superior performance but rather their ability to combine it with the right set of resources (Bharadwaj, 2000; Gregory et al., 2017; Hunt & Morgan, 1997; Trainor et al., 2014). This assertion makes the ability to identify the set of resources that integrates well with social media for exporting SMEs an important scientific enquiry.

Social Media Export Marketing Capabilities

Marketing capabilities have been studied by previous researchers (Barney, 1991; Day, 1994; Hunt, 2011) and also the issue of technology capabilities (Bharadwaj, Bharadwaj, & Konsynski, 1999; Rasiah & Myint, 2013; Wignaraja, 2002). Capabilities have been defined as 'an organisation's ability to assemble, integrate and deplore resources in combination to achieve a competitive advantage' (Trainor et al., 2014). In social media marketing, Alarcón-del-Amo et al. (2016) define social media capabilities 'as an integrated set of strategic resources and capabilities that can create, through the use of social media applications, sound competitive advantages and superior firm performance based upon a more effective information management'. This is distinct from 'an integrated set of technological resources' which is termed as social media competence (Alarcón-del-Amo et al., 2015). Some scholars have tried to explain social media internationalisation capabilities. For instance, Paniagua et al. (2016) list identity, conversation, sharing, presence, relationship, reputation and group as social media resources. They, therefore, argue that converting these social media resources into capabilities that enable a firm to internationalise will constitute social media internationalisation capabilities. The assertion, however, does not explain how the social media resource is converted into capabilities.

Based on various explanations, this chapter attempts to define social media export marketing capabilities as a firm's ability to effectively and efficiently combine social media marketing resources with key strategic firm resources to achieve superior export performance (Alarcón-del-Amo et al., 2017; Hunt & Morgan, 1997; Trainor et al., 2014). The more significant concern though is the ability to determine which firm resources combined with social media marketing can create the export capabilities. This chapter agrees with R-A theorists that resources available to firms are financial, physical, legal, human, organisational, informational and relational. It however opines that since SMEs are dominated by owner-managers (Paul et al., 2017) and are constrained financially, the owner-manager who is social media savvy is likely to be the key or even sole person updating the social media platforms, responding to customer enquiries and regularly engaging with customers, some of whom may be living in places with different time zones. As a result, an own-manager will require enough time to effectively and efficiently utilise the social media resource. Time availability, therefore, becomes an important factor in the operations of exporting SMEs. Scholars such as Klein (2007) have suggested that though some researchers consider time as a commodity, time also exhibits features that qualify it as a resource. In corroboration with the argument by Klein (2007), this chapter considers time as a resource that exporting SMEs present on social media require. Premised on the R-A theory, it is advocated that exporting SMEs would need to effectively and efficiently combine social media marketing with financial, physical, organisational, legal, human, relational, informational (market intelligence) and time to develop capabilities for export performance. This study, therefore, extends the R-A theory by proposing time as an 8th resource to the existing seven. It further argues that resources could produce different outcomes as held by the R-A competitive position matrix (Hunt, 2001; Hunt & Arnett, 2003). Some will generate a marketplace position of competitive advantage and others competitive disadvantage (Hunt, 2012). It also postulates that some resources may be more complementary compared to others. The eight resources can, therefore, be categorised into core and secondary resources. These set of firm resources can, therefore, be empirically verified to determine which ones can be considered as core, secondary or irrelevant to exporting SMEs.

Social Media and Export Relationship Performance

Export performance has been measured using financial and accounting ratios such as return on investment, return on equity and the various profitability ratios (Lages & Lages, 2004) or through the use of other absolute values such as export sales volume or export market share (Beleska-

Spasova, 2014). It has also been measured using non-financial performance indicators which are considered subjective but relevant in determining export performance (Morgan, Vorhies, & Mason, 2009). The inclusion of non-financial export performance measurements has been largely necessitated by the difficulty in gathering financial data on SMEs in particular (Ibeh & Wheeler, 2005). However social media is first of all a social online technology (Schultz & Peltier, 2013) and thrives on networking and relationship which are also essential ingredients for SME success (Felix et al., 2017; Kaplan & Haenlein, 2010; Kietzmann, Hermkens, McCarthy, & Silvestre, 2011; Nguyen et al., 2015; Trainor et al., 2014). This chapter, therefore, argues that the measurement of export performance of SMEs on social media platforms cannot ignore the firm's success at maintaining relationships with foreign customers, as one of the performance indicators. An export performance tripod is proposed with the third leg categorised relationship performance.

Although export relationship performance can be categorised as a non-financial measurement, the centrality of networking and relationship to social media marketing, requires that export relationship should be treated as the third leg of export performance measurement. Trainor et al. (2014) conceptualised relationship performance within the social media context, developed a scale and tested it. This study adopts this and terms it 'export relationship performance'. It is defined in this chapter as the level of relationship between exporting firms and their foreign customers (Trainor et al., 2014). Adapting from Trainor et al. (2014), export relationship performance measures

1. how long foreign customers have worked with firms,
2. how long new foreign customers tend to stay,
3. how loyal they perceive foreign customers to be,
4. how satisfied they perceive foreign customers to be and
5. how important customer retention is to the firm.

These five measurable variables will help to determine how successfully a firm on social media has effectively utilised the social media technology to engage customers in the export market segment. Introducing the third measurement, which has not received adequate attention in the area of export performance, is considered an important contribution to the literature on export performance.

Conceptual Framework

Following from the discourse, it is argued that SMEs can achieve export performance because they are successful at efficiently and effectively integrating the social media (platforms) resource with complementary strategic resources and aligning them to the interest of the foreign customer (Seggie & Griffith, 2008; Trainor et al., 2014). This study, therefore, proposes a framework that encapsulates the various social media platforms that exporting SMEs are leveraging on and the resources they combine to achieve export performance which includes relationship performance as demonstrated in Fig. 5.1.

Notably, WhatsApp, Facebook, YouTube, Facebook Messenger and Instagram are amongst the topmost social media platforms in the world (Genc & Oksuz, 2015). However, since Facebook messenger is an application on Facebook, this research merges it with Facebook as one platform. The study then adopts the seven resources listed by R-A theory: financial, physical, legal, human, organisational, informational and relational (Hunt, 2012), and includes 'time' as the eighth resource. The chapter posits that exporting SMEs should integrate these resources with social media, effectively and efficiently to achieve financial, non-financial and export relationship performances (Fig. 5.2).

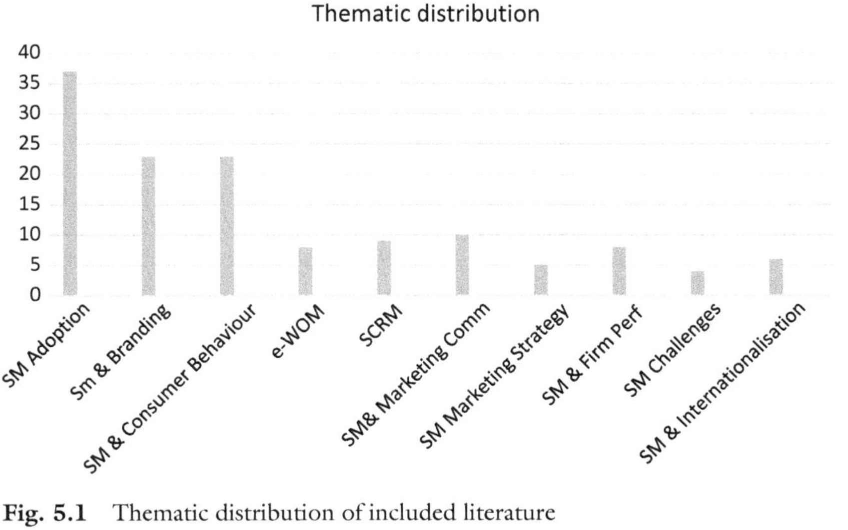

Fig. 5.1 Thematic distribution of included literature

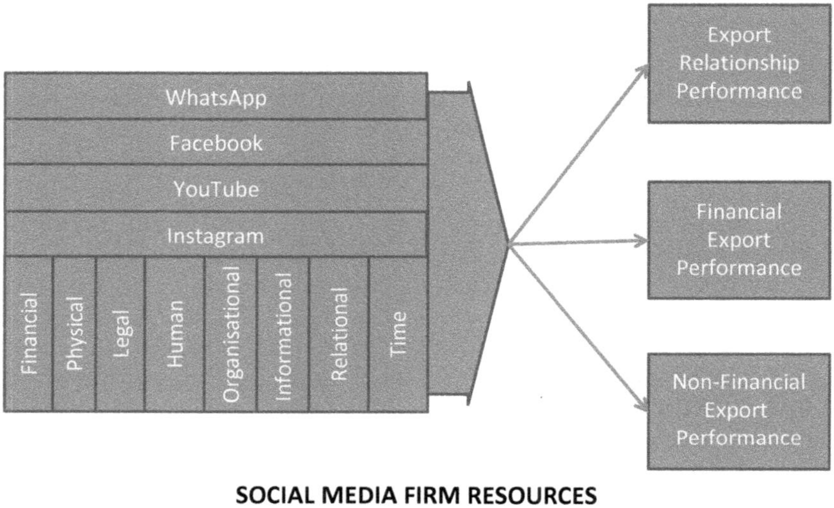

Fig. 5.2 Social media export marketing performance capabilities framework

Conclusion

Social media, a relatively cheap resource available to otherwise cash-strapped SMEs (Alarcón-del-Amo et al., 2016) is creating a new era for businesses, with the competency to improve international communication, efficiency in the international market transaction, loyalty with foreign customers and development of international network (Alarcón-del-Amo et al., 2015). It facilitates firm internationalisation by enabling firms to reach out to customers in other countries, while avoiding the cost and hurdles of physically setting up in the foreign market (Berthon et al., 2012; Okazaki & Taylor, 2013). However, research on social media marketing and export is scanty (Paniagua et al., 2016). Although few studies in the area have affirmed the positive relationship between social media marketing and export performance, there is limited focus on the combination of firm resources that enables performance.

This chapter therefore posits that since firms combine resources (Hunt, 2012), and firm performance is enhanced by the conversion of combined resources into capabilities (Alarcón-del-Amo et al., 2016; Bharadwaj, 2000; Bharadwaj et al., 1999; Trainor et al., 2014), exporting SMEs are able to achieve export performance by efficiently and effectively combining social media marketing with financial, physical, legal, organisational, human, relational and informational, and the new addition, time, resource. The chapter further suggests that firm resources are not of equal importance and therefore advocates a further qualification of the resources into core and secondary. The study also posits that to investigate social media marketing and export, export performance should not be assessed solely based on the traditional financial and non-financial indicators but also on an additional factor, export relationship performance. However, the arguments made can only be generalised if they are empirically validated. Further research is therefore proposed to test the assertions made.

References

Akar, E., & Topçu, B. (2011). An examination of the factors influencing consumers' attitudes toward social media marketing. *Journal of Internet Commerce, 10*(1), 35–67.

Alarcón-del-Amo, M. C., Rialp, A., & Rialp, J. (2015). The effect of social media adoption on exporting firms' performance. *Advances in International Marketing, 25*, 161–186.

Alarcón-del-Amo, M. C., Rialp, A., & Rialp, J. (2016). Social media adoption by exporters: The export-dependence moderating role. *Spanish Journal of Marketing—ESIC, 20*, 81–92.

Alarcón-del-Amo, M. C., Rialp, A., & Rialp, J. (2017). Examining the impact of managerial involvement with social media on Exporting firm performance. *International Business Review*. https://doi.org/10.1016/j.ibusrev.2017.09.003

Ananda, A. S., Hernández-García, A., & Lamberti, L. (2016). N-REL: A comprehensive framework of social media marketing strategic actions for marketing organizations. *Journal of Innovation & Knowledge, 1*(3), 170–180.

Barney, J. (1991). Firm resources and sustained competitive advantage. *Journal of Management, 17*(1), 99–120.

Barua, M. E., & Michie, D. A. (2014, November 4–8). The role of opportunism within the context of resource advantage theory: A conceptual framework. In Meter, R. V., & Weiser, J. (Eds.), *Advances in marketing transformational marketing: Proceedings of 2014 Society for Marketing Advances, New Orleans*. Houston: Society for Marketing Advances. Retrieved from https://www.societyformarketingadvances.org/resources/Documents/Resources/Conference%20Proceedings/2014_SMA_Proceedings.pdf#page=412.

Beleska-Spasova, E. (2014). Determinants and measures of export performance—Comprehensive literature review. *JCEBI, 1*(1), 63–74.

Berthon, P. R., Pitt, L. F., Plangger, K., & Shapiro, D. (2012). Marketing meets Web 2.0, social media, and creative consumers: Implications for international marketing strategy. *Business Horizons, 55*, 261–271.

Bharadwaj, A. S. (2000). A resource-based perspective on information technology capability and firm performance: An empirical investigation. *MIS Quarterly, 24*(1), 169–196.

Bharadwaj, A., Bharadwaj, S., & Konsynski, B. (1999). Information technology effects on firm performance as measured by Tobin's q. *Management Science, 45*(7), 1008–1024.

Boateng, R., Heeks, R., Molla, A., & Hinson, R. (2011). Advancing e-commerce beyond readiness in a developing country: Experiences of Ghanaian firms. *Journal of Electronic Commerce in Organizations, 9*(1), 1–16.

Bruhn, M., Schoenmueller, V., & Schäfer, D. B. (2012). Are social media replacing traditional media in terms of brand equity creation? *Management Research Review, 35*(9), 770–790.

Chikandiwa, S. T., Contogiannis, E., & Jembere, E. (2013). The adoption of social media marketing in South African banks. *European Business Review, 25*(4), 365–381.

Chang, Y. T., Yu, H., & Lu, H. P. (2015). Persuasive messages, popularity cohesion, and message diffusion in social media marketing. *Journal of Business Research, 68*(4), 777–782.

Compass-UPS. (n.d.). Retrieved from https://compass.ups.com/how-social-media-can-boost-exports/

Constantinides, E. (2014). Foundations of social media marketing. *Procedia—Social and Behavioral Sciences, 14*, 40–57.

Day, G. S. (1994). The capabilities of market-driven organizations. *Journal of Marketing, 58*(4), 37–52.

Elena, C. A. (2016). Social media—A strategy in developing customer relationship management. *Procedia Economics and Finance, 39*, 785–790.

Erkan, I., & Evans, C. (2016). The influence of eWOM in social media on consumers' purchase intentions: An extended approach to information adoption. *Computers in Human Behavior, 61*, 47–55.

Felix, R., Rauschnabel, P. A., & Hinsch, C. (2017). Elements of strategic social media marketing: A holistic framework. *Journal of Business Research, 70*, 118–126.

Galati, A., Crescimanno, M., Tinervian, S., & Fagnani, F. (2017). Social media as a strategic marketing tool in the Sicilian wine industry: Evidence from Facebook. *Wine Economics and Policy, 6*, 40–47.

Genc, M., & Oksuz, B. (2015). A fact or an illusion: Effective social media usage of female entrepreneurs. *Procedia—Social and Behavioral Sciences, 195*, 293–300.

Godey, B., Manthiou, A., Pederzoli, D., Rokka, J., Aiello, G., Donvito, R., & Singh, R. (2016). Social media marketing efforts of luxury brands: Influence on brand equity and consumer behavior. *Journal of Business Research, 69*(12), 5833–5841.

Gregory, G. D., Ngo, L. V., & Karavdic, M. (2017). Developing e-commerce marketing capabilities and efficiencies for enhanced performance in business-to-business export ventures. *Industrial Marketing Management*. https://doi.org/10.1016/j.indmarman.2017.03.002

Grönroos, C. (1994). From marketing mix to relationship marketing: Towards a paradigm shift in marketing. *Management Decision, 32*(2), 4–20.

Hinson, R. (2010). The value chain and e-business in exporting: Case studies from Ghana's non-traditional export (NTE) sector. *Telematics and Informatics, 27*, 323–340.

Hu, Y.-C., & Wang, F. C.-C. (2009). Collectivism, corporate social responsibility, and resource advantages in retailing. *Journal of Business Ethics, 86*(1), 1–13.

Hunt, S. D. (2001). Commentary—A general theory of competition: Issues, answers and an invitation. *European Journal of Marketing, 35*(5/6), 524–548.

Hunt, S. D. (2011). Developing successful theories in marketing: Insights from resource-advantage theory. *AMS Review, 1*, 72–84.

Hunt, S. D. (2012). The evolution of resource-advantage theory: Six events, six realizations, six contributions. *Journal of Historical Research in Marketing, 4*(1), 7–29.

Hunt, S. D., & Arnett, D. B. (2003). Resource-advantage theory and embeddedness: Explaining R-A theory's explanatory success. *Journal of Marketing Theory and Practice, 11*(1), 1–17.

Hunt, S. D., & Morgan, R. M. (1995). The comparative advantage theory of competition. *Journal of Marketing, 59*, 1–15.

Hunt, S. D., & Morgan, R. M. (1997). Resource-advantage theory: A snake swallowing its tail or a general theory of competition? *Journal of Marketing, 61*, 74–82.

Ibeh, K. I. N., & Wheeler, C. N. (2005). A resource-centred interpretation of export performance. *International Entrepreneurship and Management Journal, 1*, 539–556.

Kaplan, A. M., & Haenlein, M. (2010). Users of the world, unite! The challenges and opportunities of social media. *Business Horizons, 53*(1), 59–68.

Kietzmann, J. H., Hermkens, K., McCarthy, I. P., & Silvestre, B. S. (2011). Social media? Get serious! Understanding the functional building blocks of social media. *Business Horizons, 54*, 241–251.

Killian, G., & McManus, K. (2015). A marketing communications approach for the digital era: Managerial guidelines for social media integration. *Business Horizons, 58*, 539–549.

Kim, E., & Drumwright, M. (2016). Engaging consumers and building relationships in social media: How social relatedness influences intrinsic vs. extrinsic consumer motivation. *Computer Human Behaviour, 63*, 970–979.

Klein, C. C. (2007). The Economics of Time as a resource. Retrieved from http://citeseerx.ist.psu.edu/viewdoc/download?doi=10.1.1.329.7498&rep=rep1&type=pdf.

Lages, L. F., & Lages, C. R. (2004). A measure of short-term export performance improvement. *Journal of International Marketing, 12*(1), 36–56.

Lu, V. N., & Julian, C. C. (2008). The internet, strategy and performance: A study of Australian export market ventures. *Journal of Global Marketing, 21*(3), 231–240.

Luo, X., Zhang, J., & Duan, W. (2013). Social media and firm equity value. *Information Systems Research, 24*(1), 146–163.

Malthouse, E. C., Haenlein, M., Skiera, B., Wege, E., & Zhang, M. (2013). Managing customer relationships in the social media era: Introducing the social CRM house. *Journal of Interactive Marketing, 27*(4), 270–280.

Mangold, W. G., & Faulds, D. J. (2009). Social media: The new hybrid element of the promotion mix. *Business Horizons, 52*(4), 357–365.

Martin, J., Goncalves, R., Oliveira, T., Cota, M., & Branco, F. (2016). Understanding the determinants of social network sites adoption at firm level: A mixed methodology approach. *Electronic Commerce Research and Applications, 18*, 10–26.

Mehrtens, J., Cragg, P. B., & Mills, A. M. (2001). A model of Internet adoption by SMEs. *Information & Management, 39*, 165–176.

Morgan, N. A., Vorhies, D. W., & Mason, C. H. (2009). Market orientation, marketing capabilities, and firm performance. *Strategic Management Journal, 30*, 909–920.

Moro, S., Rita, P., & Vala, B. (2016). Predicting social media performance metrics and evaluation of the impact on brand building: A data mining approach. *Journal of Business Research, 69*(9), 3341–3351.

Nguyen, B., Yu, X., Melewar, T. C., & Chen, J. (2015). Brand innovation and social media: Knowledge acquisition from social media, market orientation, and the moderating role of social media strategic capability. *Industrial Marketing Management, 51*, 11–25.

Odoom, R., Anning-Dorson, T., & Acheampong, G. (2017). Antecedents of social media usage and performance benefits in small- and medium-sized enterprises (SMEs). *Journal of Enterprise Information Management, 30*(3), 383–399.

Okazaki, S., & Taylor, C. R. (2013). Social media and international advertising: Theoretical challenges and future directions. *International Marketing Review, 30*(1), 56–71.

Opoku, R. A., Abratt, R., Bendixen, M., & Pitt, L. (2007). Communicating brand personality: Are the web sites doing the talking for food SMEs? *Qualitative Market Research: An International Journal, 10*(4), 362–374.

O'Reilly, T. (2007). *What is Web 2.0: Design patterns and business models for the next generation of software*. Retrieved from http://www.oreilly.com/pub/a/web2/archive/what-is-web-20.html.

Packer, R. (2011). Social media marketing: The art of conversational sales. Retrieved from https://wsimarketbuilders.com/socialmediamarketing-whitepaper.pdf.

Paniagua, J., Korzynski, P., & Mas-Tur, A. (2016). Crossing borders with social media: Online social networks and FDI. *European Management Journal, 35*(3), 314–326.

Paniagua, J., & Sapena, J. (2014). Business performance and social media: Love or hate? *Business Horizons, 57*, 719–728.

Paul, J., Parthasarathy, S., & Gupta, P. (2017). Exporting challenges of SMEs: A review and future research agenda. *Journal of World Business, 52*, 327–342.

Rasiah, R., & Myint, M. M. (2013). Ownership, technological capabilities and exports of garment firms in Myanmar. *Technological and Economic Development of Economy, 19*(1), S22–S42.

Rest, J.-P., & Roper, A. (2013). A resource-advantage perspective on pricing: Shifting the focus from ends to means-end in pricing research? *Journal of Strategic Marketing, 21*(6), 484–498.

Schlegelmilch, B. B. (2011). Commentary on developing successful theories in marketing: Insights from resource-advantage theory. *Academy of Marketing Science Review, 1*, 85–89.

Schultz, D. E., & Peltier, J. (2013). Social media's slippery slope: Challenges, opportunities and future research directions. *Journal of Research in Interactive Marketing, 7*(2), 86–99.

Seggie, S. H., & Griffith, D. A. (2008). The resource matching foundations of competitive advantage: An alternative perspective on the globalization of service firms. *International Marketing Review, 25*(3), 262–275.

Sinkovics, N., Sinkovics, R. R., & Jean, R. B. (2013). The internet as an alternative path to internationalization? *International Marketing Review, 30*(2), 130–155.

Socialpilot.co. (n.d.). Retrieved May 3, 2017, from https://socialpilot.co/blog/125-amazing-social-media-statistics-know-2016/.

Stoian, M. C., Rialp, A., & Rialp, J. (2011). Export performance under the microscope: A glance through Spanish lenses. *International Business Review, 20*(2), 117–135.

Teng, S., Khong, K. W., Chong, A. Y. L., & Lin, B. (2017). Persuasive electronic word-of-mouth messages in social media. *Journal of Computer. Information Systems, 57*(1), 76–88.

Thoeni, A. T., Marshall, G. W., & Campbell, S. M. (2016). A resource-advantage theory typology of strategic segmentation. *European Journal of Marketing, 50*(12), 2192–2215.

Trainor, K. J., Andzulis, J. M., Rapp, A., & Agnihotri, R. (2014). Social media technology usage and customer relationship performance: A capabilities-based examination of social CRM. *Journal of Business Research, 67*(6), 1201–1208.

Tuten, T. L., & Solomon, M. R. (2015). *Social media marketing* (2nd ed.). Sage.

Wang, W. Y. C., Pauleen, D. J., & Zhang, T. (2016). How social media applications affect B2B communication and improve business performance in SMEs. *Industrial Marketing Management, 54*, 4–14.

Wang, Z., & Kim, H. G. (2017). Can social media marketing improve customer relationship capabilities and firm performance? Dynamic capability perspective. *Journal of Interactive Marketing, 39*, 15–26.

Wignaraja, G. (2002). Firm size, technological capabilities and market-oriented policies in Mauritius. *Oxford Development Studies, 30*(1), 87–104.

Yadav, M., & Rahman, Z. (2017). Measuring consumer perception of social media marketing activities in e-commerce industry: Scale development & validation. *Telematics and Informatics, 34*, 1294–1307.

Zhu, Y-Q., & Chen, H-G. (2015). Social media and human need satisfaction: Implications for social media marketing. *Business Horizons, 58*, 335–345.

CHAPTER 6

Data Analytics and Predictive Analytics: How Technology Fits into the Equation

Brian J. Galli and Gabrielle Muniz

Introduction

Throughout the years, technology has changed considerably. In the mid-1960s, mainframe computers were state of the art in the information-processing kingdom. Although IBM has existed since 1952, behemoths were produced by the 1960s and 1970s. The term "mainframe computer" was largely synonymous with IBM because of the company's dominant market share at the time. These things were room-filling metal contraptions of vacuum tubes and wires that had to be air-cooled to avoid overheating. The cost was upwards of $3 million in 1960s dollars (Klimberg, 2016)! Also, memory was about 5121 KB, and hard drive storage was about 350 MB, while processing speed was a scorching 1 MIPS (million instructions per second). This, of course, mirrors Bill Gates's (in)famous 1981 statement: "640 KB ought to be enough for anybody."

However, there was quite a reversal by 2016 with the iPhone 6. This phone featured 32 GB (+9300%) of memory and was pocket sized. Author and futurist Ray Kurtzweil notes that cell phones are much cheaper, more

B. J. Galli (✉) • G. Muniz
Long Island University, Brookville, NY, USA
e-mail: Brian.Galli@liu.edu

B. George, J. Paul (eds.), *Digital Transformation in Business and Society*, https://doi.org/10.1007/978-3-030-08277-2_6

powerful, and more compact than a standard computer. Compared to the computer at Massachusetts Institute of Technology (MIT) in 1965, the cell phone is a billion times more capable per dollar/euro (Klimberg, 2016). Though most corporate executives would suggest that their two most valuable corporate assets are their people (employees) and their products, a closer look at the products and their composition reveals that data is often listed as a primary component. That would include customer data and the actual product data required for creating output. Many businesses have now come to embrace this concept and have built analytics or intelligence units to better understand and utilize the data within their control (Buchanan, 2017).

According to IBM, there are many advanced analytics capabilities that come from predictive analytics, for example ad hoc statistical analysis, real-time scoring, predictive modeling, text analytics, and data mining, to name only a few. With these tools, patterns can be found in data to help businesses to predict the future. The field of "big data" has also expanded to include activities around data mining, machine learning, and artificial intelligence to better optimize the "10,000-foot view" of a company's business. The term "predictive analytics" can be thought of as a catch-all for disciplines, which includes regression analysis: a technique that looks at values of correlated variables to prove or disprove a specific set of assumptions (Marvin, 2016).

Furthermore, predictive analytics is about recognizing patterns in data to project probability, according to Allison Snow, senior analyst of business-to-business (B2B) marketing at Forrester. One must realize that analytics is centered on probabilities, rather than facts. It is impossible to know what data is significant with predictive analytics, while it is usually the opposite case for traditional analytics. With predictive analytics, one can indicate the predictive data for the particular outcome that is being predicted (Marvin, 2016). For example, a company may have a model for predicting how much a particular customer may purchase over time. The model could also be based on purchases that the customer has already made, which predicts the next product that they are likely to buy. Sales forecasting and the use of marketing models to plan where best to place advertising buys are examples of predictive analytics (Davenport, 2014).

Thomas H. Davenport, a distinguished professor in management and information technology at Babson College, is a research fellow at the MIT

Initiative on the Digital Economy. He observes that the three components of supporting predictive analysis—the data, the statistics, and the assumptions—are all critical contributors to the final product. They are compared to the legs of a three-legged stool, each of which can cause a collapse if it is not functioning correctly. Davenport points out that the assumptions underlying a data set or statistical evaluation process are equally important to the data and the statistics. The primary assumption in a predictive analysis is often that the future always mirrors the past. For instance, the financial crisis between 2008 and 2009 was the result of invalid models that predicted the likelihood of mortgage customers repaying their loans, since the decline of house prices was not considered. As house prices fell, the models were useless in predicting mortgage repayment, given that a hidden assumption was that housing prices would always rise (Davenport, 2014). Clearly, the data and the statistics provide a necessary balance for conclusions drawn from the assumptions.

Moreover, both data analytics and predictive analytics are potent tools for extrapolating conclusions from data. Data analytics and predictive analytics are usually associated with being effective means of information analysis that exchange privacy for results. Also, there are staggering numbers of applications for this kind of analysis, especially with the various methods of recording information. Many companies are already applying these analytical techniques, and they will continue to refine them as the field expands. As more technology is being incorporated into products, those products provide more data for the producing companies. When data is made more readily available, then companies can find correlations between trends based on that data. Thus, they can draw different conclusions from the premises of their research.

Though data analytics and predictive analytics can be used to learn many things, they are not without their downsides. To reveal these downsides, we will explore both current and future applications of data analytics and predictive analytics, as well as how they function in different environments. We will explore how the flexible nature of the fields allows for data analytics and predictive analytics to apply to many purposes and contexts. Afterwards, we will explore the various benefits and drawbacks of using data analytics. Understanding how this type of analytics will be used, what is required, and the results that it provides will help to reveal which fields will most benefit from this kind of analytics.

The Value of Analytics

Essentially, there are a vast number of opinions about the usefulness and reliability of data and predictive analytics, since data analytics is applicable to many different fields. The uses of data analytics are innumerable, but more often than not they will relate back to theory testing. For example, someone will generate a theory based on existing data, a hypothesis, or even a whim, and will then create a model to process data to find correlations between that theory and the accumulated data. There is a great amount of data within scientific fields (i.e., medicine, biology, physics, etc.), so big data analytics can open the door for many scientific breakthroughs (Zhao, Gao, & Liu, 2018).

Regarding how the data is being processed, it is somewhat different for comprehensive data analytics and predictive analytics. Data analytics requires the manipulation of a data set for information, either about the evidence itself or a specific conclusion. Information about evidence is typically referred to as metadata, and it helps to observe trends, patterns, and so on. Predictive analytics works on a different level, while also working from a base data set. It uses both data and metadata to determine the likelihood of an event. Furthermore, predictive analytics is an excellent technique for determining risk management, but its accuracy is determined both by the data size and by the accuracy of existing data. Subsequently, we will be observing the uses of data and predictive analytics from medicinal, governmental, business, personal, scholastic, and financial perspectives.

The highest concentration of predictive analytics appears to be located in the pharmaceutical world, at least from the pool of information that can be observed. Healthcare professionals can make use of these techniques to determine which medications are right for individual patients, as models can forecast the risks and benefits of using medications (Flynn & Stevenson, 2018). This would require large companies to share data and existing predictive models, such as with the US Food and Drugs Administration's (FDA) Sentinel Initiative. Flynn and Stevenson also discuss the potential of wearable technology to give medication providers valuable data, as well as the tradeoff that many users may not want their information to be shared with those companies (Flynn & Stevenson, 2018). Trading privacy for better-tailored products, whether it is medicine or something else, is a common theme with these analytics (Manisha & Lathwal, 2016).

On the other hand, government institutions, such as the US Internal Revenue Service (IRS), make use of data analysis, but it is done in a somewhat dubious manner. While data mining often goes hand in hand with government bodies, such as the National Security Agency (NSA) or the Department of Defense (DOD), it can be used to determine a variety of different things for other than security purposes. The IRS has moved on to creating profiles of taxpayers that are constructed from both public and commercial data sets, with the data processed using analytics. However, Houser and Sanders argue that these practices work against fair information practices and are a substantial threat to privacy. While these practices are very cost efficient for the IRS, they are invasive of taxpayer privacy and can lead to further government overreach (Houser & Sanders, 2017).

Additionally, data and predictive analytics can be applied in the workplace environment. Tasks such as hiring, training, and employee performance can also be enhanced by data analytics. Many companies use data analytics to make observations about their environment, but rarely do they use them to improve their training and business development. A business can analyze its data to figure out more cost-effective means of operating, as well as which hires are more likely to be successful (Giacumo & Breman, 2016).

Another newer method of data collection is coming from automobiles, specifically from electric vehicles. Analysts are referring to the information that is collected from these cars to make future improvements to their design without using any personal information about the consumer. Various aspects are recorded, such as charging time, charger style, driving distances, daily drives, charging frequencies, and other information about product use. By observing the typical usage patterns for their vehicles, designers can design better and newer versions of those products to cater to the data collected (Ahmed & Kapadia, 2017).

Outside of a business environment, predictive analytics can be applied to personal uses: identifying and preventing spam e-mails, for instance. Due to the variable nature of spam, it is tough to filter out completely. A system that can analyze e-mail logs and make use of predictive analytics can build trust criteria to determine which senders are considered spam (Vesely, 2017). Examining previous instances of spam would allow for the system to recognize spam behavior and to act accordingly when e-mails that fit the behavior profile are sent. Such a system could update its behavioral patterns based on newly formed spam (Yuksel, Cankaya, & Uncu, 2017).

In the scholastic setting, predictive analytics used analytical techniques to measure how likely students were to succeed. The documented system was referred to as Predictive Analytics Reporting or PAR. The method developed could be applied to show a student's likelihood of passing a class, but Wagner and Longanecker make clear that the framework was designed with both conventional and non-conventional students in mind. Most federal data collection only focused on full-time students, which led to an incomplete data set. PAR was created to analyze the unchanging rates of postsecondary education success, as they have not changed for the past 40 years, despite attempts being made to improve them (Wagner & Longanecker, 2016).

Lastly, predictive analysis has various applications in a financial setting. John Blossom calls the multiple devices that collect and analyze data the "signal economy." More specifically, this is an economy that is driven by analysis of the data collected by those numerous different devices. In other words, it is significant data that has been processed to pinpoint trends. By subjecting economic data to the same processes that one may run to find the likeliness of an event's occurrence, one can figure out all sorts of potential areas of growth. Due to the staggering amount of data available from all kinds of different sources, markets, news outlets, universities, and even social media, software is required to process all of that data. Such data is continuously being collected and handled by software for hedge funds, the government, or some other private institution. Clearly, the information that comes from this kind of inquiry can predict how markets will act if enough data is being appropriately processed (Blossom, 2014).

As one can see, there are many applications for data analytics and predictive analytics that span pretty much every field. Many institutions, both public and private, make use of these analytical practices, and these processes are improving over time. In its current state, predictive analytics is more or less a means of trying to go as close as possible to determine the likelihood of a future event. The more data that is used, the more probability there will be of precision. With the advent of technological convergence, more devices can record information which will make it possible to map out the conditions for any event. From that point, analyzing the states that surround events will lead to the development of more accurate predictive analysis tools.

How Technology Impacts Analytics

Evidently, these analytics have proven to be beneficial in many aspects of human life. Predicting the future is a coveted tool in one's lifetime, especially within organizations. No one could dislike the ability to foresee success when creating a product, as investments could be much more easily made with little to no error. With these tools, many types of success could be guaranteed.

However, there can be a cost to such powerful tools if they are not used responsibly. In Thomas Davenport's foreword for Eric Siegel's book *Predictive Analytics: The Power to Predict Who Will Click, Buy, Lie or Die*, he observes that the developing science of predictive analytics has, by and large, been for the good of humanity. Davenport asserts that predictive analytics can save lives, such as with healthcare, crime, and terrorism. He also addresses how these approaches can be detrimental, as one must exercise responsibility so as not to abuse this power. Large amounts of data would be required from people and companies, so predictive models can be compared to other powerful technologies or disruptive human innovations (Davenport, 2014). Davenport further suggests that Siegel counterpoints Nassim Nicholas Taleb's *The Black Swan* and other works. Taleb posits that most predictive efforts must be doomed to fail from the inherent unpredictability of things, but Siegel is the first to acknowledge that black swan events occur beyond prediction (Siegel, 2013). Furthermore, potential business applications continue to multiply. Models range from how predictive analytics is changing the retail industry to fintech start-ups using predictive modeling on fraud analysis and financial transaction risk. In essence, we have not fully tapped into the ways for industries to integrate data analysis and how predictive analytics tools can redefine business methods. After the breakthrough of mapping an artificial brain through creating a computer, there are infinite possibilities (Marvin, 2016).

Applications of Technology in Analytics

There are numerous examples of predictive analytics applications in multiple industries, especially involving technology. Primarily, marketing intersects with many sectors, as marketing campaigns have become much more scientific through the use of predictive analytics and big data. The ability to predict and recommend future purchases and actions is the principal benefit for many companies, and predictive analytics and big data are gateways to such a benefit.

With these tools, customer satisfaction and network safety can both be improved for some major corporations. For example, one of the largest users of predictive analytics is Amazon. The website uses past purchases and known "likes" to identify, predict, and offer other items to a customer. Similarly, fraud detection is a field where analyzing past patterns is key in detecting future threats or criminal activity. Cybersecurity has become a huge budget item at many large corporations, such as banks, insurance companies, investment firms, credit card issuers, social media companies, and even law enforcement agencies (Moghaddass, Zuo, Liu, & Huang, 2015). PayPal avoids network breaches by using historical payment data, device type, country of origin, and user profiles to develop algorithms that detect fraud during any transaction. Even insurance companies, such as Liberty Mutual, use predictive analytics techniques to help evaluate lifestyle, health, and other factors to calculate life expectancy and related premiums. Furthermore, airlines use providers to analyze dynamic fare pricing that is based on the predicted number of travelers over a given period, for instance Virgin Atlantic utilizes Amadeus. Hotels do the same with room rates based on anticipated future travel volume (Farooq, 2016). Clearly, predictive analytics can help many corporations to gain a competitive edge.

By studying the company called Endor, one can find that *social physics* can also provide a massive amount of data in a matter of minutes. The science of social physics uses machine learning and algorithmic models to understand and predict crowd behaviors. It is comparable to Google, as it provides simple and fast results for any question (Matheson, 2017). Endor was recently spun out of the MIT Media Lab incubator, and it has developed a new platform that allows businesses to ask real-time questions: "Where should we open our next store?" or "Who is likely to try product X?" In one recent exercise, Endor analyzed Twitter data for an unnamed US defense agency to detect potential terrorist activity. Endor gathered 15 million data points with examples of 50 Twitter accounts of known ISIS activists that were based on identifiers in the metadata. The challenge was to detect new accounts with identifiers extremely well hidden in the metadata. Someone at Endor completed the task on a laptop in 24 minutes, detecting 80 "lookalike" ISIS accounts, 45 of which were from the pool of well-hidden accounts that were suspected by the agency (Ghofrani, He, Goverde, & Liu, 2018). As one can see, social physics was able to provide helpful results in a matter of minutes.

Endor is becoming a well-known company that is making great strides with predictive analytics. It was co-founded by Alex "Sandy" Pentland, Toshiba Professor of Media Arts and Sciences, and Yaniv Altshuler, a former MIT postdoctoral researcher. Early testing on the platform was done with the US Defense Advanced Research Project Agency (DARPA). It analyzed mobile data from individual cities in times of civil unrest to show how emerging patterns can help predict future riots. Lastly, Endor recently won "Cool Vendor" status from Gartner, reserved for industry disrupters, and was acknowledged as a "Technological Pioneer" by the World Economic Forum (Matheson, 2017). It is evident that the merging of analytics and technology can guarantee more opportunities than one could have ever imagined over the history of industrialization.

How Technology and Analytics Come Together

The two types of analytics considered in this chapter, data and predictive, are powerful means for analyzing data in any number of different fields. Evidently, the results of the analyses can provide information to keep people safe, to make people money, to save people time, and so forth. Unfortunately, there is a cost in relation to privacy, as large amounts of data have to be analyzed to attain such results. With smart technology and smart environments, more data can be collected and processed, which will increase the results from data analytics and predictive analytics. These are great tools that are sure to grow and evolve with their related industries because they are derived from data. The immense applicability of these analytical methods ensures that they will be around for quite some time. Overall, it will be exciting to see how these analytical methods are applied to both new and existing fields, as it is inevitable that even more areas will jump on the bandwagon.

References

Ahmed, N. K., & Kapadia, J. (2017). Big data analytics: How big data is shaping our understanding of electrified vehicle customers. *SAE International Journal of Materials & Manufacturing, 12*, 99–107.

Blossom, J. (2014). The Signal Economy: How to target new revenues through predictive analytics. *Information Services & Use, 34*, 17–25.

Buchanan, E. (2017). Considering the ethics of significant data research: A case of Twitter and ISIS/ISIL. *PLoS ONE, 12*, 1–6.

Davenport, T. H. (2014). A predictive analytics primer. *Harvard Business Review*.

Farooq, M. (2016). Applications of predictive analytics in various industries. *BigData-MadeSimple.com*. Retrieved from http://bigdata-madesimple.com/applications-of-predictive-analytics-in-various-industries-2/

Flynn, A. J., & Stevenson, J. G. (2018). The future of data, analytics, and information technology. *Pharmacy Forecast*, pp. 31–34.

Ghofrani, F., He, Q., Goverde, R. M., & Liu, X. (2018). Recent applications of big data analytics in railway transportation systems: A survey. *Transportation Research: Part C, 90*, 226–246.

Giacumo, L. A., & Breman, J. (2016). Emerging evidence on the use of big data and analytics in workplace learning. *Quarterly Review of Distance Education, 17*, 21–38.

Houser, K. A., & Sanders, D. (2017). The use of big data analytics by the IRS: Efficient solution or the end of piracy as we know it? *Vanderbilt Journal of Entertainment & Technology Law, 14*, 817–872.

Klimberg, R. K. (2016). *Fundamentals of predictive analytics with JMP, Second edition*. Cary, NC: SAS Institute.

Manisha, A., & Lathwal, P. (2016). Exploring classification & clustering techniques for predictive analytics. *International Journal of Recent Research Aspects, 6*, 76–78.

Marvin, R. (2016). Predictive analytics, big data, and how to make them work for you. *PC Magazine*.

Matheson, R. (2017, December 19). Inventing the "Google" for predictive analytics. *MIT News*. Retrieved from http://news.mit.edu/2017/endor-inventing-google-predictive-analytics-1220

Moghaddass, R., Zuo, M., Liu, Y., & Huang, H.-Z. (2015). Predictive analytics using a nonhomogenous semi-Markov model and inspection data. *IIE Transactions, 47*, 505–520.

Siegel, E. (2013). *Predictive analytics: The power to predict who will click, buy, lie or die*. Hoboken, NJ: John Wiley & Sons, Inc.

Vesely, R. (2017). Predictive analytics: IU Health knows the patient in Room 103 is at high-risk for CLABSI (cover story). *H&HN: Hospitals & Health Networks, 91*, 20–25.

Wagner, E., & Longanecker, D. (2016). Scaling student success with predictive analytics: Reflections after four years in the data trenches. *Change, 48*, 52–59.

Yuksel, A. S., Cankaya, S. F., & Uncu, I. S. (2017). Design of a machine learning predictive analytics system for spam Problem. *Acta Physica Polonica, A, 132*, 500–504.

Zhao, W., Gao, L., & Liu, A. (2018). Programming foundations for scientific big data analytics. *Scientific Programming, 5*, 1–2.

CHAPTER 7

Effective Engagement of Digital Natives in the Ever-Transforming Digital World

Anju Varghese Philip and Zakkariya K. A.

Introduction

According to the latest estimates, more than 4.2 billion people worldwide have access to the Internet (Internet World Stats, 2018). That number is more than half of the world's present population. The digital divide is being demolished every second of every day and we are fast moving into a hyper-connected networked world where past boundaries of geographical distances, time zones, and cultural differences are ceasing to matter. The average person uses more than four connected devices every day, and digital technology has arrived as the invisible but all-knowing, all-controlling dimension of our everyday lives.

Modern-day businesses have to thrive in a data-driven culture where the only thing that does not change is change itself. And the speed at which the changes happen is increasing so exponentially that it feels like everyone is on a constant learning loop. Jumping off this loop is equivalent to committing professional suicide as you can no longer refuse to embrace constant learning and change. How can businesses negotiate a world where everything is digital and connected and your customers are at

A. V. Philip (✉) • Zakkariya K. A.
Cochin University of Science and Technology, Kochi, India

B. George, J. Paul (eds.), *Digital Transformation in Business and Society*, https://doi.org/10.1007/978-3-030-08277-2_7

times more informed than you are? The best solution is by letting go of the scary connotations of the phrase 'digital disruption' and embracing the possibilities and opportunities of 'digital transformation.'

Digital transformation can be defined as the acceleration of business activities, processes, competencies, and models to fully leverage the changes and opportunities of digital technologies and their impact in a strategic and prioritized way (Edmead, 2016). It can also be called digital business transformation, since digital transformation is in essence transformation of the underlying framework of your business too. Technology innovation, customer behavior and demand, and external environmental factors are some of the reasons why a business might want to undergo a digital business transformation (Edmead, 2016). For this we need to build 'digital fluency' among our employees as well as increase the 'digital readiness' of our organizations (Briggs & Makice, 2012).

Digital fluency is an ability to reliably achieve desired outcomes through use of technology. It is the maximum individual potential to achieve desired outcomes through the use of digital technology (Briggs & Makice, 2012). It is not enough to know how to use the technological tools but one has to understand how to use them to achieve the desired outcomes. Therefore, being digitally fluent is a step up from mere digital literacy. A digitally literate person will have the basic skills required to participate in the digitally enabled social sphere, but, in contrast, a digitally fluent person will be able to navigate the same with an ease that comes from constant practice and immersion in the technology. Companies can be described as being 'digitally ready' when they are able to support the digital fluency of their employees by aligning their resources, culture, and purpose with the tasks they are asked to undertake (Briggs & Makice, 2012). Digital fluency is not a static state of being. It is a state of constant learning, up-gradation, and up-skilling because of the speed with which digital technologies change.

Like for any language, being digitally fluent implies that you have a level of comfort with the digital world that implies that you were born speaking the digital language. And that is where the digital natives or millennials enter the discussion. Who are the digital natives? Why are they important in a discussion about digital transformation of business? Digital natives (Prensky, 2001) is the terminology used to describe that part of the world's population who were born after 1980 and have been immersed in a technological environment all their lives, thereby making them especially adept at handling changing technology with an almost effortless flair.

Digital immigrants, on the other hand, lived part of their lives in a simpler time where technology was not the driving force of their lives and hence have a more difficult time navigating in and adapting to today's technology led milieu. It is expected that by 2025, digital natives will comprise 75% of the workforce, thereby making them the most economically powerful demographic group in history (dmresourcecenter.com, 2018).

In this scenario, it is important for businesses to understand who the digital natives are so that they can better leverage them as employees and also understand what makes them tick as customers. An organization that has to manage a successful digital business transformation will require a workforce that consists of both digital natives and digital immigrants to cooperate with the process and successfully make the transition so that no one is left feeling overwhelmed and falls behind in the process. This chapter attempts to throw some light on how to manage digital natives both as internal and as external stakeholders of the business.

Getting Familiar with the Digital Natives

To quote Mark Prensky (2001), digital natives are '"native speakers" of the digital language of computers, video games and the Internet.' 'Digital immigrants' are those who need to learn the digital language, they are not born speaking it, so they carry an 'accent' no matter how proficient they become at using digital tools. Digital immigrants prefer to use e-mail for online communication, whereas digital natives prefer the more synchronous forms of instant messaging. With mobile phones, digital immigrants favor speaking to people, whereas digital natives prefer speed texting. Digital natives prefer to share differently as well. Blogging is increasingly becoming popular among both digital immigrants and digital natives, but once again for different reasons. Digital natives use blogging to share personal experiences and treat personal blogging websites as forms of online journals. By contrast, digital immigrants tend to use blogging sites more as intellectual tools to share and discuss ideas with their peers (Prensky, 2001).

Not only do digital natives use technology differently—their lives are also being impacted by technology differently to the extent that their self-presentation, self-image, self-esteem, and wellbeing are influenced (Colbert, Yee, & George, 2016). They are often described as digitally literate, highly connected, experiential, social, and in need of immediate gratification (McMahon & Pospisil, 2005). Digital natives consider the digital world to be part of their personalities. They expect to be connected at all times (Vodanovich, Sundaram, & Myers, 2010).

Digital is almost coded into the genes of the millennial generation; it is that natural for them. They are digital by default. Research has shown that they spend 41% of their free time on their devices (Morrison, 2015). They own an average of 2.4 devices, which might include smartphones, laptops, notebooks, iPads, and music streaming gadgets. Prensky (2001) says that brain scans show that their increased use of digital devices has actually caused changes to their brain structure. They relate to technology in a different way and are naturals at interacting and understanding with technology.

Among the adverse consequences of this constant connectedness is what is called FOMO in urban slang—the Fear Of Missing Out. It is a compulsive need to be up to date in what is going on within their social circles as well as in the outer world, which can even lead to disruption of healthy sleep patterns as people take their smartphones to bed with them and check messages and other updates during the nights and first thing in the morning.

At the same time, many among digital natives say they feel a sense of social fatigue, which is 'a feeling that being connected all the time is too much. A desire to disconnect and be offline and to regulate and bring down the time spent on social media' (techopedia.com). As a result, the more conscientious among them try to limit their time spent on social media by prioritizing their social media use and limiting themselves to being active on only a handful of social media sites rather than trying to be everywhere all at once and getting overwhelmed by it.

Many digital natives have grown up playing video games and have acquired a set of unique skills such as the ability to take risks and overcome obstacles, to learn from failures, to manage diverse teams, solve conflict, distribute rewards equitably, and so on as a result (Colbert et al., 2016). These skills make them invaluable assets in the workplace. Similarly, businesses that are looking to implement 'design thinking' for their innovation process will do well to inculcate the gaming skills of learning by trial and error (Beck & Wade, 2004; Glen, Suciu, & Baughn, 2014).

The digital workforce will likely be comfortable with technology-based instruction (Kraiger & Ford, 2006), giving organizations a low-cost, replicable solution for helping employees develop the skills that they need. Some tools that are currently in use are Google Drive for collaborative writing, Trello for collaborative project management, and Yammer or

Slack for communication and enterprise-level social networking. With more organizations using firm-generated social media content to connect with customers and build their brands, employees who understand and can leverage the power of social media will also be valuable to organizations (Kumar, Bezawada, Rishika, Janakiraman, & Kannan, 2016).

On the other hand, many millennials experience a decrease in face-to-face communication skills and nervousness during telephone calls because most communication is conducted via texting which is seen as more efficient and gives the opportunity to self-present in a favorable manner. Also they display decreasing levels of empathy probably because of increasing levels of technology immersion (Colbert et al., 2016).

Technology has blurred the lines between work and nonwork domains (Ramarajan & Reid, 2013). While email, the Internet, and even social media are integral tools for doing work, they also provide easy access to family, friends, online shopping, and other nonwork purposes while at work. Inability to disconnect from technology during nonwork hours can lead to higher levels of stress and work-life conflict (Ramarajan & Reid, 2013) adversely affecting employee wellbeing (Colbert et al., 2016).

What Can Companies Do for Digital Natives?

As Employees

Digital natives are highly cause driven and will love to work for companies that measure performance not just in financial terms but also in terms of contribution to social good. They are also aware of and support brands that have a social cause (dmresourcecenter.com, 2018).

While digital natives boast a strong work ethic, how, when, and where they accomplish the work should be left to them. Companies that expect a 9–5, 5-day workweek from them can expect a dissatisfied bunch of employees (Lovelock, 2018). Work-life balance is very important to them, especially because they have the technological tools that make this balance easier to achieve. While they want to earn enough to facilitate their life goals, they also want the free time to pursue their interests outside of work (Ramarajan & Reid, 2013).

They have grown up in a culture where collaboration is important (Commonplaces, 2014) and companies should provide the necessary internal knowledge management tools to accomplish this. They seek opinions and inputs from various quarters and believe that this gives them better solutions to problems. Companies should be open to setting up collaborative teams not limited by location or department. Virtual teams are a great way to accomplish this (Gilson, Maynard, Jones Young, Vartiainen, & Hakonen, 2015).

Digital natives are not intimidated by titles, seniority, or positions. They see their superiors as mentors rather than bosses and will expect to be treated as equals. They are highly ambitious and expect to make managerial positions in two years of joining the work force (Ashridge and MSLGroup, 2014).

Loyalty to a particular organization is not important to them and they will switch jobs if they get bored. Millennials will not stay in a company and wait for a promotion every 3–5 years (Ashridge and MSLGroup, 2014). Companies should function as corporate lattices where employees are moved horizontally across business functions for different projects rather than as corporate ladders where the only movement is vertical (Benko, Anderson, & Vickberg, 2011).

Mobility is an important factor for millennials (Gilson et al., 2015). Companies should provide the necessary technological support to enable this. A combination of cloud computing and use of light technological solutions to enable mobility will increase efficiency in the workplace too. Also an option to bring your own technology (BYOT) will be a boon to this workforce (Puybaraud, 2012).

As Customers

For marketers and brands seeking to engage their millennial audience, it is first of all imperative to realize that as a result of living all their lives in an environment where information is available at their fingertips, they are impossible to fool and incredibly impatient with slow responses and campaigns that don't really speak to them. They carry their idealism proudly and will happily purchase brands and products that espouse a social cause. They are also vociferous in speaking out about anything that they dislike, so companies need to be very careful about how they engage with this audience online (dmresourcecenter.com, 2018). Be quick at apologizing for any mistakes and try to rectify them at the earliest. Best strategy will be to not make any mistakes at all!

Millennials are also characterized by their focus on accumulating life experiences rather than owning property. So they might opt to use their income to travel rather than settling down into a typical middle-class lifestyle buying homes, cars, and other large immovable investments

(Niewiadomski & Anderson, 2015). This demography is moving from a lifestyle of ownership to a lifestyle of easy access, instead of buying a car they will use a ride aggregator such as Uber, instead of investing on office space, they might prefer to work from shared office spaces. This type of thinking gives them all the conveniences of modern life that money can provide access to without having to bother with the hassles that come with ownership such as maintenance and upkeep.

They are constantly digitally networked at all times (Sørensen, 2016), which means they have easy access to information at all times and this means that they are attuned to an on demand lifestyle and have a growing sense of entitlement. Millennials handle half their ecommerce transactions from a mobile device and are the largest online audience and have more buying power than any other previous generation (dmresourcecenter.com, 2018).

Therefore, given their high buying power, their influence on the online market, their active social media presence, and most of all their impatience with delay, it is imperative that brands make all effort to understand and engage them. This is especially pertinent in a country like India, which is expected to become the youngest country in the world by 2022, with the average age of 29 years (Financial Express, 2017).

Firms should wake up to the advantages of using social media as a business tool to engage their digitally savvy audience (Lovelock, 2018). An effective social media marketing strategy can help increase an organization's brand awareness, improve customer engagement and satisfaction, and allow the company to establish their voice. Social media provides an accessible platform for direct communication between a company and a client and allows for a more human-centric approach to customer success and support. Firm-generated content (FGC) has a greater effect on customers with a longer customer–firm relationship, who are technologically savvy, and who are more prone to social networking. It is social media's ability to give 'voice' to customers in the form of likes and comments that makes FGC more effective. FGC can play a key role in strengthening customers' relationship with the firm by encouraging them to buy across several product categories (Kumar et al., 2016).

Statistics show that desktop sales are falling rapidly, while smartphones are being used for everything ranging from work to leisure. So the way to engage the consumer of today is by going mobile and social and being entertaining while doing so. That is spend more on mobile and digital campaigns. Give your target consumers a chance to connect with their friends and other like-minded people through your campaign.

Use the trails of digital data that is left on the Internet by your consumers to really understand them. Social Media Analytics and Big Data should be seen as an opportunity to understand consumers and customize and individualize your offerings rather than as an insurmountable challenge best left to the geeks. While designing platforms for customer engagement remember to personalize, interact, be intuitive, attractive, and social (Vodanovich et al., 2010). Companies can work on enhancing the customer experience through understanding customers by using digital trails, analytics-driven customer segmentation, and digitally enhanced selling.

Skills That Are Needed for the Digital Workplace

Many times the obstacle toward implementing a successful digital transformation is people and not technology (Bonnet & Nandan, 2011). This is because digitalization changes the ways of working and increases the rate of change of the organization. It will require the manpower to rapidly acquire new skills and competencies, and also usher in a new organizational culture. Leadership also has to change to new forms to accommodate and enable this transformation to digital. Digital transformation requires the continuous alignment of people, technologies, organizational structures, and ultimately organizational culture (Briggs & Makice, 2012). Effective change management is required to manage this process well and adopting a design thinking style will help with the implementation (Gruber, De Leon, George, & Thompson, 2015). Design thinking can be defined as 'a human-centered approach to innovation that puts observation and discovery of often highly nuanced, even tacit human needs right at the forefront of the innovation process. It considers not just the technological system constraints but the socio-cultural system context' (Glen et al., 2014).

The speed and complexity of digitalization will change how we handle information (find, filter, and use information from many sources), interact with people (choosing an identity or identities to engage with others through multiple channels), inspire, and involve people to act in support of organizational goals, and empower them to move toward a future reality. To succeed in the digital age, a person must be able to appropriately weight the practical risks and future benefits of new tech. New levels of skilling required to make the transition to digital fluency transcend from basic computer skills such as conducting a basic search for information, creating a profile on different social media apps to performing basic troubleshooting tasks to more advanced ones such as editing and uploading photos, tagging content, crafting semi-public content for various contexts, navigating complex privacy controls, and editing a wiki page. Some of the skills exhibited by an email-able person remain useful: good typing skills, writing a persuasive argument, and finding an optimal time to communicate.

Adaptation is what allows us to use our abilities effectively when shifting between contexts, usually as the world changes around us. Adaptation is an *ability of abilities*, drawing from the knowledge and relevant skills one accumulated from previous experience. Adaptation requires a mindset that sees all of this change as an opportunity to learn.

Briggs and Makice (2012) listed the skills and abilities needed for employees to make a successful digital transition as enumerated by Jenkins (2006) and summarized them as given below:

1. *Networking*: The ability to search for, synthesize, and disseminate information
2. *Play*: The capacity to experiment with one's surroundings as a form of problem-solving
3. *Performance*: The ability to adopt alternative identities for the purpose of improvisation and discovery
4. *Simulation*: The ability to interpret and construct dynamic models for real-world processes
5. *Appropriation*: The ability to meaningfully sample and remix media content
6. *Multitasking*: The ability to scan one's environment and shift focus as needed to salient details
7. *Distributed cognition*: The ability to interact meaningfully with tools that expand mental capacities
8. *Collective intelligence*: The ability to pool knowledge and compare notes with others toward a common goal
9. *Judgement*: The ability to evaluate the reliability and credibility of different information sources
10. *Transmedia navigation*: The ability to follow the flow of stories and information across multiple modalities

11. *Negotiation*: The ability to travel across diverse communities, discerning and respecting multiple perspectives, and grasping and following alternative norms
12. *Visualization*: The ability to interpret and create data representations for the purposes of expressing ideas, finding patterns, and identifying trends

A number of tools are now available to measure the new skills needed. These include the iSkills assessment (Katz, 2007), New Literacies (Lankshear & Knobel, 2006), and the New Media Literacies (NML) framework (Literat, 2014).

Discussion and Conclusion

Digital natives are a great mascot for organizational transformation programs, as they have the ability to come into a company unhindered by previous experience with legacy technologies and provide new solutions to modernize business processes and drive a truly digital workplace. As digital natives join the workforce, they bring with them a willingness to embrace technology as a driver of not just productivity, but also profitability. Organizations must adopt this mindset and plan for digital disruption in the modern workplace, or risk being left behind.

Digital natives can help their older colleagues use technology more efficiently and effectively, thereby increasing productivity. The technology gap between generations will lessen as elder generations learn to embrace and understand new technologies like their digital native peers. Digital natives' higher level of intuition when it comes to technology, coupled with the fact that new products are easier to use than their predecessors, means that integration with technology will be less of an issue for all generations (Gilson et al., 2015). In addition, digital natives have a healthy level of respect for the older generations' work ethic and values and will want to imbibe these from their seniors at work.

Ultimately, digital natives seek the same things from their employer as any other generation that came before them. A sense of belonging and value and an opportunity to contribute to the success of their employer. They seek to be respected and nurtured and to work for companies that work for social causes and innovation and have a sense of social responsibility. They want their work and lives to be an extension of their own identities and will want to work in and buy from companies that provide them this avenue for self-expression.

As customers, digital natives are highly tech savvy, value conscious, and cause driven. They desire instant gratification and quick responses for their issues. Companies should set in place efficient teams to monitor, engage, and address their millennial customers, as this is considered the most economically powerful demographic that the world has ever seen. The most effective customer touch point for this audience would be some form of social media outreach in addition to the usual sales and promotion activities. Companies that go the extra mile to really understand their digital native market will be able to target their offerings in a way that truly speaks to the identity of the millennial.

To conclude, business leaders should acknowledge the direction that technology is leading the world in, embrace the future, and ride the waves of technology into the hearts and minds of their digital native stakeholders. Since the number of digital native stakeholders are continuously increasing and leading to a world filled only with digital natives in the long run, companies should have short-term and long-term plans and strategies for attracting, engaging, and managing digital native to retain them as loyal stakeholders. Digital transformation is a never-ending process as the technology is growing and transforming every day, and this poses a challenge to organizations, as they need to cater to the ever-changing needs and behaviors of the digital native. No company would be able to continuously either engage or satisfy the digital native without having continuously upgraded technologically based plans, strategies, and actions.

References

Ashridge and MSLGroup. (2014). The Millennial Compass—Truths about the 30-and-under generation in the workplace. Retrieved from http://www.scribd.com/doc/211602632/The-Millennial-Compass-The-MillennialGeneration-In-The-Workplace.

Beck, J. C., & Wade, M. (2004). *Got game: How the gamer generation is reshaping business forever*. Boston, MA: Harvard Business School Press.

Benko, C., Anderson, M., & Vickberg, S. (2011). The corporate lattice: A strategic response to the changing world of work. *Deloitte Review*, Issue 8

Bonnet, D., & Nandan, P. (2011). *Transform to the power of digital: Digital transformation as a driver of corporate performance* (Report). Capgemini Consulting.

Briggs, C., & Makice, K. (2012). *Digital fluency: Building success in the digital age*. SociaLens.

Colbert, A., Yee, N., & George, G. (2016). The digital workforce and the workplace of the future. *Academy of Management Journal, 59*(3), 731–739.

Commonplaces. (2014, January 21). Digital natives and the future of the workplace [Blog post]. Retrieved October 10, 2018, from https://www.commonplaces.com/blog/digital-natives-and-the-future-of-the-workplace/

Edmead, M. (2016, May 2). *Digital transformation: Why it's important to your organization*. Retrieved October 13, 2018, from https://www.cio.com/article/3063620/it-strategy/digital-transformation-why-its-important-to-your-organization.html

Gilson, L. L., Maynard, M. T., Jones Young, N. C., Vartiainen, M., & Hakonen, M. (2015). Virtual teams research: 10 years, 10 themes, and 10 opportunities. *Journal of Management, 41*(5), 1313–1337.

Glen, R., Suciu, C., & Baughn, C. (2014). The need for design thinking in business schools. *Academy of Management Learning & Education, 13*(4), 653–667.

Gruber, M., De Leon, N., George, G., & Thompson, P. (2015). From the editors: Managing by design. *Academy of Management Journal, 58*(1), 1–7.

Internet World Stats: Usage and population statistics. (2018, June 30). Retrieved October 13, 2018, from https://www.internetworldstats.com/stats.htm.

Jenkins, H. (2006). *Confronting the challenges of participatory culture: Media education for the 21st century*. White paper for MacArthur Foundation.

Katz, I. R. (2007). Testing information literacy in digital environments: ETS's iSkills assessment. *Information Technology and Libraries, 26*(3), 3–12.

Kraiger, K., & Ford, J. K. (2006). The expanding role of workplace training: Themes and trends influencing training research and practice. In L. L. Koppes (Ed.), *Historical perspectives in industrial and organizational psychology* (pp. 281–309). Mahwah, NJ: Lawrence Erlbaum Associates Publishers.

Kumar, A., Bezawada, R., Rishika, R., Janakiraman, R., & Kannan, P. K. (2016). From social to sale: The effects of firm-generated content in social media on customer behavior. *Journal of Marketing, 80*(1), 7–25.

Lankshear, C., & Knobel, M. (2006). *New literacies: Everyday practices and classroom learning*. Open University Press.

Literat, I. (2014). Measuring new media literacies: Towards the development of a comprehensive assessment tool. *Journal of Media Literacy Education, 6*(1), 15–27.

Lovelock, D. (2018, August 14). *How digital natives are changing the modern workplace*. Retrieved October 13, 2018, from https://www.nintex.com/blog/how-digital-natives-are-changing-the-modern-workplace/

McMahon, M., & Pospisil, R. (2005). Laptops for a digital lifestyle: Millennial students and wireless mobile technologies. *Proceedings of the Australasian Society for Computers in Learning in Tertiary Education, 2*, 421–431.

Morrison, K. (2015, August 11). *Report: Digital natives do everything from mobile devices*. Retrieved October 11, 2018, from https://www.adweek.com/digital/report-digital-natives-do-everything-from-mobile-devices/

Niewiadomski, R., & Anderson, D. (2015). *Digital Generation: Is this the beginning of paradigm shift in ownership?* Retrieved October 13, 2018, from https://cie.acm.org/articles/digital-generation-beginning-paradigm-shift-ownership/.

Prensky, M. (2001). Digital natives, digital immigrants part 1. *On the Horizon, 9*(5), 1–6.

Puybaraud, M. (2012, February 1). *Digital natives: A tech-savvy generation enters the workplace.* Retrieved October 14, 2018, from https://workdesign.com/2012/02/digital-natives-a-tech-savvy-generation-enters-the-workplace/

Ramarajan, L., & Reid, E. (2013). Shattering the myth of separate worlds: Negotiating nonwork identities at work. *Academy of Management Review, 38*(4), 621–644.

Sørensen, C. (2016). The curse of the smart machine? Digitalisation and the children of the mainframe. *Scandinavian Journal of Information Systems, 28*(2), 57–68.

The impact of digital natives on the future of business. (2018). Retrieved October 11, 2018, from http://dmresourcecenter.com/unit-1/digital-natives/

Vodanovich, S., Sundaram, D., & Myers, M. (2010). Research commentary—Digital natives and ubiquitous information systems. *Information Systems Research, 21*(4), 711–723.

'*With an average age of 29, India will be the world's youngest country by 2020*'. (2017, March 26). Retrieved October 14, 2018, from https://www.financialexpress.com/india-news/with-an-average-age-of-29-india-will-be-the-worlds-youngest-country-by-2020/603435/

CHAPTER 8

Robotics, Artificial Intelligence, and the Evolving Nature of Work

Craig Webster and Stanislav Ivanov

Introduction

In their thought-provoking paper Brynjolffson and McAfee (2015) ask the question 'Will humans go the way of horses?' and they have good ground to do so. Human labour has been largely replaced by machines in ways that consumers can understand and this has been going on for decades. For example, many will have a historical memory of going to a bank during the bank's hours of operation to make a withdrawal of cash. Now, consumers regularly go to machines to make cash withdrawals from the accounts and the labour for making the transaction is largely mechanized. While much work in the service industry is still based upon human labour, manufacturing is much more automated. While humans may still be needed to perform many commercial operations, human labour is much less critical than it had been at any time before in history.

C. Webster (✉)
Ball State University, Muncie, IN, USA
e-mail: cwebster3@bsu.edu

S. Ivanov
Varna University of Management, Varna, Bulgaria
e-mail: stanislav.ivanov@vumk.eu

B. George, J. Paul (eds.), *Digital Transformation in Business and Society*, https://doi.org/10.1007/978-3-030-08277-2_8

Recent advances in robotics, artificial intelligence, and automation technologies (RAIA) (Bhaumik, 2018; Miller & Miller, 2017; Neapolitan & Jiang, 2013; Russell & Norvig, 2016; Warwick, 2012) have allowed companies from various sectors of the economy to adopt RAIA in search of lower costs, faster production time, provision of consistent product quality, management of supply chain operations, etc. While initially it was the manufacturing that used industrial robots (Colestock, 2005; Cubero, 2007; Low, 2007; Pires, 2007), currently all other sectors of the economy and society intensively adopt RAIA technologies: from supply chain management (Min, 2010), agriculture (Driessen & Heutinck, 2015), autonomous vehicles (Maurer, Gerdes, Lenz, & Winner, 2016), and warfare (Crootof, 2015; Sparrow, 2007), through travel, tourism, and hospitality (Ivanov & Webster, 2018; Ivanov, Webster, & Berezina, 2017; Murphy, Hofacker, & Gretzel, 2017), education (Ivanov, 2016; Timms, 2016), journalism (Clerwall, 2014), provision of legal (Remus & Levy, 2015), and other services (Huang & Rust, 2018; Wirtz et al., 2018), to trading on the financial markets (Dunis, Middleton, Karathanasopolous, & Theofilatos, 2017), and implementing medical operations (Kaur, 2012; Mirheydar & Parsons, 2013; Schommer, Patel, Mouraviev, Thomas, & Thiel, 2017). Both large and small companies use chatbots to communicate and maintain their relationship with customers (Hill, Ford, & Farreras, 2015; Xu, Liu, Guo, Sinha, & Akkiraju, 2017). Social robots enter actively our lives (Agah, Cabibihan, Howard, Salichs, & He, 2016; Nørskov, 2016; Royakkers & van Est, 2016), and already redefine our understanding of 'sex' (Cheok, Devlin, & Levy, 2017; Lee, 2017). These brief examples reveal the overwhelming intertwining of RAIA technologies into the fabric of society, leading to huge transformations of the ways humans live, work, and do business (Makridakis, 2017; Talwar, 2015; Talwar, Wells, Whittington, Koury, & Romero, 2017). Some authors even state that economists and marketers need to broaden their definition of a 'consumer' to include robots in it (Ivanov & Webster, 2017).

The advances in RAIA technologies are expected to accelerate in the future and some researchers consider the possibility that artificial intelligence might at one point surpass human intelligence—a point of time usually referred to as 'technological singularity' (Callaghan, Miller, Yampolskiy, & Armstrong, 2017; Kurzweil, 2005; Shanahan, 2015). While we do not know whether silicon will surpass carbon in terms of intelligences and whether transhumanism and human augmentation through microchip brain implants is a viable future and not science fiction, we are sure that technological advances in RAIA, Industry 4.0 (Andelfinger & Hänisch, 2017; Schwab, 2016; Skilton & Hovsepian, 2018), and the Internet of things (Sendler, 2018) are challenging the role of human labour as a production factor to a point when all or an overwhelming share of goods and services are delivered by RAIA technologies instead of human employees. Such an economic system is called 'robonomics' (Crews, 2016; Ivanov, 2017). Experts' opinions about that automated robot-based economy and society are not uniform. One group of researchers expresses explicit fear of AI technologies (Barrat, 2013; Bostrum, 2014; Crews, 2016; Leonhard, 2016), considers the self-aware AI as 'our final invention' (Barrat, 2013) and paints dystopian dooms-day Terminator-like scenarios for the human species. Others seem overly optimistic and perceive technology as the ultimate utopian solution to all human problems, including the merger between humans and machines (Kurzweil, 2005). The majority of authors, however, take a more pragmatic approach and see the opportunities that AI and robonomics as an economic system would create like extended life expectancy, improved health and quality of life, more time for activities people value, to name just a few (Brynjolfsson & McAfee, 2014; Frank, Roehring, & Pring, 2017; LaGrandeur & Hughes, 2017; Talwar, 2015; Talwar et al., 2017), although the expected radical abundance of goods and services that technology experts promise (Drexler, 2013) may not benefit all stakeholders.

Within this domain of thought, the future of human's work and the impacts of digitalization and automation on the labour market have received a great deal of attention by researchers (e.g., Alasoini, 2017; Association for Advancing Automation, 2017; Daugherty & Wilson, 2018; Ford, 2009, 2015; Frank et al., 2017; Harteis, 2018; Kaplan, 2015; Kauhanen, 2016; West, 2018). As a matter of fact, in a recent highly cited research, Frey and Osborne (2017) assess that 47% of US jobs are already susceptible to computerization. Some jobs are very obviously on the brink of extinction. For example, the most common job in 29 of the 50 states in the United States is the 'truck driver,' according to 2014 data (NPR, 2015) and the self-driving car/truck (a technology on the brink of being viable on the road) will almost immediately lead to the unemployment of these workers (Solon, 2016).

This raises justified concerns among researchers, politicians, and industry representatives how people, companies, economies, governments, and societies as a whole would need to adapt to the new technological, economic, social, and political realities that robonomics would create. For example, in 2016, the Executive Office of the President of the United States, National Science and Technology Council, and the Committee on Technology published three documents intended to prepare the US economy for the age of automation: *The National Artificial Intelligence Research and Development Strategic Plan* (2016), *Preparing for The Future of Artificial Intelligence* (2016), and *Artificial Intelligence, Automation, and the Economy* (2016). In the same manner, in 2017, OECD published an extensive report dedicated to the implications of automation for businesses and governments (OECD, 2017), while the World Economic Forum published annual reports on the future of jobs (WEF, 2016, 2018).

This chapter contributes to the body of knowledge by elaborating on the impacts of RAIA technologies on the labour market and the nature of work. This chapter looks into such things as the employment dilemma (Giarini & Liedtke, 1996), an issue related to the human need to have productive work and the modern industrial capability for productivity. It also looks into the issue of the human Luddite possibilities, as humans should have a fear of being replaced by robots (Lehman, 2015; McClure, 2017). In addition, the chapter also looks into the issue of the ethical use of robots (Lin, Abney, & Bekey, 2014), as robots are not merely replacements of human labour but also entails a real impact upon the human population.

The Economics of Robotics and AI

The first element to look into is the issue of why RAIA technologies are being implemented. The obvious reasons for the implementation are the overlap between the technological abilities that now exist and the need for efficiency and cost-effective production. While the concept of the robot was invented in 1920 (NPR, 2011), the robot (and more importantly the intelligent robot) was a thing of the future and not entirely practical. By the 1980s, robots were being utilized widely in manufacturing, although service industries have generally lagged behind in the use of robots.

At its heart, the current robotic economy undermines the economic principle of the labour theory of value was ascribed to by such thinkers as Karl Marx, Adam Smith, and David Ricardo. While traditional economic theory identifies land, labour, and capital (and sometimes also 'enterprise' or 'entrepreneurship' as the factors of production), we see a shift in the stress on which of these factors is most critical. Traditional economic theory holds that human labour is a necessary factor for the production of wealth, meaning that human labour is critical in the productive process, something that any casual reading of Marx makes very clear. However, with the rapid advances in robotics and artificial intelligence, human labour is much less stressed as a necessary input, and this will be increasingly true once robots are used to produce robots.

The widespread implementation of a robotic economy will lead to a new economic and social landscape in which there will be winners and losers. Since the technological capability to replace human labour with increasingly sophisticated technologies exists, those humans with labour to sell that can be replaced easily and cheaply with new technologies will be replaced, leading to a new political economy of robots (Kiggins, 2018). This will be a new world in which most labour in terms of production of goods and services is created by machines, meaning that humans will be increasingly marginalized from the productive process, although there is the enduring question of whether robots will largely be used as complements to human labour or substitutes to human labour (DeCanio, 2016).

Another political and economic trend that will impact upon the new economy is a by-product of the globalized economy. Since economies are now interconnected as never before and there are massive advantages to those who conquer a market early, since markets tend to be big and the ability to penetrate such markets. What we can expect in markets is the markets will increasingly gravitate towards 'winner-take-all' markets (Brynjolfsson & McAfee, 2014). There are a number of reasons for this, including the digitalization, improvements in communications (and transportation), and the increased importance of networks and standards. In other words, the economy is now different, as global markets enable consumers to have access to information with regards to goods and services and information costs with regards to products and transportation of products is dropped, and networks enable the spread of information and good and services as never before. Economies of scale develop, enabling the establishment of near monopolies, as penetrating markets outside of prevailing channels and networks becomes increasingly difficult for start-ups.

The value of labour and the way that labour is treated can be understood by its value in the market. The relationship between increased labour productivity and wages in the United States tell an interesting story. While the data in the United States show massive increases in productivity per worker since the 1950s, by the mid-1970s, increases in salaries have not kept up with increases in productivity (Watson, 2018). It seems that companies see no need to compensate workers for their productivity, resulting in an outcome that has a massive impact upon the economy and society, people who are employed and productive but do not experience salary/wage increases as a result of their labour. It is also suggestive that workers are so productive that there are underemployed populations driving the real wages down. What is most interesting is that this is occurring in economies in which there are demographic declines, meaning that the available labour pool is generally shrinking and will continue to shrink. Another issue is the question of the sustainability of such economies, as machines do not consume many goods nor services, unlike humans (Ford, 2015, p. 196). This means that economies may suffer in the future, as machines do not consume the way that humans do. Manufacturers may be producing high-quality goods at a low price per unit, but the question remains who will be able to purchase the goods and services, when salaries remain relatively flat in many countries.

Directions of RAIA Impacts on the Labour Market

For us human mortals living today, we will have to contend with the massive changes that robotics and artificial intelligence will do to the labour market in the next 10, 20, or more years. While we understand that in 100 years, the production and service landscape will change dramatically because the capability of technologies that we now are impressed with will seem very primitive, we have to prepare for the nearer future. In the nearer future, we can expect some major impacts that are almost predictable, such as the automation of a great deal of the transportation with self-driving trucks and cars. At any rate, there will be an impact upon different countries in different ways based upon the country's current level of development and social/technological factors. There will also be some impact upon social structures and the mobility of populations, as well as the mobility of production. At any rate, legislation will have to change a great deal and the legal/political elements as well.

The first and major impact of technologies will be the elimination of jobs via automation (Frank et al., 2017), although there will be many jobs lost, there will also be the creation of new jobs and the changing face of other jobs. Some publications have discussed various solutions to how technological unemployment can be dealt with (Stevens & Marchant, 2017; Swan, 2017). Many of the jobs we now see exist in the market will be changed a great deal with many jobs being de-skilled, re-skilled, or upskilled in terms of their tasks associated with them. There is also a geographical element to this, as well. There is an indication that automation of many tasks may enable production to take place in places in which labour costs are notoriously high such as Germany, where shoe production is now taking place (Wiener, 2017) presumably because the other factors of production are equal to or lower than in Asia and labour costs approach zero because of automation. Production may return to the Western economies, as labour costs will largely be bypassed and production may now be done in ways that lower transportation costs.

Another major impact that will occur is the quality of the jobs that persist. Humans will likely continue to do some of the work that robots and computers will not yet be able to do and some of the jobs are rather surprising. For example, while we can reasonably expect that artificial intelligence can handle most of service complaints and inquiries from humans, most robots have a hard time doing some activities that many humans find very simple, such as opening doors. It seems that there will be a need for humans to do many of the physical tasks that machines cannot do or at least do well, although it is possible that social changes may work in ways to bypass this. For example, while currently no robots can change the sheets on a bed well, it is possible that hotels in the future could hand guests clean sheets and insist that the guests make their own beds, bypassing the need to pay for human labour to change the sheets on the bed. Such a social/cultural shift may be accepted by hotel guests, if there are cost advantages.

Because of the disruption of advanced robotics and artificial intelligence into the production of goods and services, humans will likely have to find a niche in the world's gig economy. Since human labour will be increasingly relegated to physical or mental tasks that humans can still perform better than machines, humans will have to hustle, doing jobs that machines cannot yet do well nor as effectively as machines. One of the problems that will be a political outcome of this is that many people will be reliant upon short-term tasks and jobs that may exist below the radar of tax authorities and do not supply the human with the legal and social protections that a person would have with a full-time job that exists above the board. How males and females will work under these conditions and survive under such conditions will be interesting, but it is so far unknown how the sexes will either react or flourish under such unstable and shifting work environments (Schwab, 2016).

The class differences will increase, as has been the case in nearly all developed countries in recent decades, since labour will consist of skilled knowledge-based workers and less-skilled workers. Those that are less skilled will constantly run into competition with robots and artificial intelligence that will undermine the value of labour (Economic Times, 2018). There will be some specializations that remain difficult to replace with robots or artificial intelligence, and these positions will continue to function, although there will be downward pressures upon such jobs, as the demand remains equal, the supply of labour for such jobs will decrease the returns for those working such jobs. However, those who feel secure with skills and are knowledge workers should not feel too secure, as there are technological changes that can make their labour redundant. For example, university professors now teach many online courses. Such courses are generally done for the convenience of the students, but there are risks for the teachers, as many courses may rely on exams/tests that are multiple choice and can already be graded with very basic computer technology, and it may not be long before computers can grade and give intelligent feedback to students on essays (see also Ivanov, 2016).

We can also expect that the disruptions that will take place will have a massive reaction among the public and there should be political responses. Since humans may be less key to the production of goods and services and many humans will be replaced with robotic/AI labour, what should be done to sustain the economy and sustain the political economy? A likely answer spoken about is guaranteeing humans a universal basic income (Santens, 2017; Sheahen, 2012). This income would have the stabilizing impact by ensuring that human consumers can purchase the goods and services that a largely automated economy would produce. Paying for such a plan would require a different form of taxation from how taxation is now considered, as a great deal of taxation in economies stems from the taxation of the wages of working people. So with a reduction in the number of humans working and paying taxes on their incomes and the reduction of jobs, entire new systems of taxation will have to be created. The current economy and its laws presume human labourers paying taxes upon their salaries and presumes that such human labour is necessary to sustain economies. Laws also presume that people are engaged in full-time employment in which people have stable incomes and benefits, meaning that many of the laws that now exist regulating labour presume an old-fashioned industrial model of labour. But the new economy, to a large extent, depends upon gigs and short-term contracts and laws have not kept up with these changes.

Staying Competitive on the Labour Market

The future, at least the near future, is one that requires a skilled workforce. Table 8.1 lists the top ten job skills the World Economic Forum (2018) feels the workforce should have in 2020. The list of skills seems to be uniquely intellectual and emotional in nature. There is no mention of physical strength or physical agility. According to the World Economic Forum (2018), the skills needed are intellectual and emotional in nature, including solving complex problems, critical thinking, and coordinating with others.

These key skills identified by the World Economic Forum presumes that the competitive advantage that humans have over robots and artificial intelligence is not based upon physicality but upon intellectual capabilities and the ability to interpret the emotional landscape of the workplace. The development of such skills would reasonably thought to be developed by people through education/training, experience, and naturally occurring intellectual ability. What this implies is that education will have to work in ways to cultivate those talents that only humans at this point in time can develop in ways that are useful in the economy (creativity, complex communication skills, emotional intelligence, etc.). The need to further develop these very human capabilities will have to depend upon an educational system that is based upon an old industrial model from a previous century, presuming a tiered system of education that would build skills that are permanent. The new economy will likely need a new approach to education, one that is based upon the development of human skills rather than building of knowledge and will have to retool to consider the needs

Table 8.1 Top ten job skills needed in 2020

Rank	*Skill*
1.	Complex Problem Solving
2.	Critical Thinking
3.	Creativity
4.	People Management
5.	Coordinating with Others
6.	Emotional Intelligence
7.	Judgement and Decision Making
8.	Service Orientation
9.	Negotiation
10.	Cognitive Flexibility

Source: World Economic Forum (2018)

of people in the workforce who will have to pick up new skills to be relevant in a flexible and changing economy. Since many employees will change jobs and industries, they will need relevant education that will enable them to have relevant skills in the new positions that they take in different companies and fields. Companies will also have to pick up some of the responsibility for the training and retraining of employees and such responsibility should not be the sole responsibility of universities, entities comfortably insulated from the economy in many respects.

Apart from how individuals and how the educational systems will have to change, industry will also have to change. In the short run, for companies to stay relevant and competitive, companies will have to invest in robots and artificial intelligence. While such investments may require investments upfront, such investments will assist in the transition to the new economy, despite the risks and hiccups in the system. The changeover for most industries is a sure thing. While adopting some technologies later may offer some advantages, adopting early may enable the customers and workforce to adjust to the new realities and work in ways that are more effective and efficient.

Conclusion

Humans have been on the Earth for quite some time and we hope that the species will persist. While humans probably will not go the way of the horses (Brynjolffson & McAfee, 2015), there are massive changes caused by technological advances that humans need to be aware of and must adjust to. First, we expect that humans will be removed increasingly from the production of goods and services. Human labour will remain one the factors of production but is decreasingly the key component in the production process. The qualities that the human labourer has and brings to the table in the factors of production will remain the ability to create, to interpret, and to work with others. The human will also be prized in the near future for the physical ability to do things that many robots have a hard time doing well and quickly (opening doors, walking, and other physical things), the human will remain a great multi-purpose tool that can negotiate the geography of a house or office space. But the flooding of labour markets with labour displaced by robots and artificial intelligence will mean that such physical jobs will be flooded.

For humans, those lacking the skills to compete in the new economy will be competing against each other. Skills that so far are unique to humans will have to be developed for many workers to remain relevant. Problem solving, emotional intelligence, interpersonal communications, and other skills that humans can do well if they are sophisticated enough to do it will enable humans to remain relevant in the new economy. But the traditional establishments that train humans will have to adjust to the needs of the new economy and the demands upon workers.

Second, the political and legal system has to take account of the new economy. As labour becomes more flexible and concentrated upon gigs and contracts, governments and taxation systems have to keep up with the new economic reality. From the perspective of the government, the traditional way of taxing working people will be less prevalent and the persistence of the gig economy/sharing economy will mean that taxation systems will have to be amended to remain relevant in an economy that has changed. Taxation will change and keeping the new economy stable and healthy will have to rely on different thinking. With a large number of workers who are not needed, guaranteeing that people have a reliable way of living, especially if they are not needed in the workforce, is a necessity. One easy way to adjust for this is the guarantee of a minimal income for all, an idea that may seem radical but has a long history and has become quite popular to speak about in recent years (see, for example, Caputo, 2012; Stern, 2016). The stability of the political economy rests upon the need to have consumers who use products and services and the lack of the need for much of the labour available in the economy means that much of the workforce will be competing for de-skilled jobs and with people who have lost their jobs due to increasing robotization of the workplace.

Finally, there is the major concern about the assumptions of the human's ability to be able to be relevant in the new economy. While the World Economic Forum (2018) can identify skills needed in the workplace, it is a question as to whether all humans have the ability to develop skills that are needed by industry. While there are some who may contend that IQ is a meaningful concept (Gould, 1996), most agree that there is variation in each human's ability to absorb, retain, and process information. Unless it is found that all humans have roughly the same intellectual capacity to develop the skills needed for the new economy, it would seem that many would be excluded from the new economy, simply because of their incapacity to develop the skills needed. It could be, then, that a substantial portion of humanity could be excluded from the new economy, simply because of a lack of the ability to develop the skills needed to compete as a human in the economy.

In conclusion, humans are entering a new age, an age of mechanical reproduction that is unlike anything envisioned before. While humans may remain a factor of production, they may be decreasingly critical in terms of the terms of physicality of work and even more critical in terms of the intellectual needs of the workplace. At any rate, the role of the human will be different. As a result, many of the institutions that are part of the economy, such as government and education, will have to change to be relevant and consistent with the needs of a society that has an increasingly automated productive base.

References

Agah, A., Cabibihan, J. J., Howard, A., Salichs, M. A., & He, H. (Eds.). (2016). *Proceedings of the 8th International Conference on Social Robotics (ICSR) 2016*, Kansas City, MO, USA, November 1–3, 2016. (Vol. 9979). Springer.

Alasoini, T. (Ed.). (2017). *Nordic Future of Work Conference I: The Future of Work and new Forms of Work from the Global and the Nordic Perspectives.* TemaNord 2017:551. Nordic Council of Ministers. https://doi.org/10.6027/09086692

Andelfinger, V. P., & Hänisch, T. (Eds.). (2017). *Industrie 4.0: Wie cyber-physische Systeme die Arbeitswelt verändern.* Wiesbaden: Springer Gabler.

Artificial Intelligence, Automation, and the Economy. (2016). Published by the Executive Office of the President of the USA. Retrieved from https://obamawhitehouse.archives.gov/sites/whitehouse.gov/files/documents/Artificial-Intelligence-Automation-Economy.PDF

Association for Advancing Automation. (2017). *Work in the automation age: Sustainable careers today and into the future.* Retrieved from http://www.a3automate.org/docs/Work-in-the-Automation-Age-White-Paper.pdf

Barrat, J. (2013). *Our final invention: Artificial intelligence and the end of the human era.* New York: Macmillan.

Bhaumik, A. (2018). *From AI to robotics: Mobile, social, and sentient robots.* Boca Raton, FL: CRC Press.

Bostrum, N. (2014). *Superintelligence: Paths, dangers, strategies.* Oxford University Press.

Brynjolfsson, E., & McAfee, A. (2014). *The second machine age: Work, progress, and prosperity in a time of brilliant technologies.* W. W. Norton & Company.

Brynjolffson, E., & McAfee, A. (2015). Will humans go the way of horses? Labor in the second machine age. *Foreign Affairs, 94*(4), 8–14.

Callaghan, V., Miller, J., Yampolskiy, R., & Armstrong, S. (Eds.). (2017). *The technological singularity: Managing the journey.* Berlin: Springer.

Caputo, R. (Ed.). (2012). *Basic income guarantee and politics: International experiences and perspectives on the viability of income guarantee.* Springer.

Cheok, A. D., Devlin, K., & Levy, D. (Eds.) (2017). Revised selected papers of the *Second International Conference on Love and Sex with Robots (LSR) 2016*, London, UK, December 19–20, 2016. Springer.

Clerwall, C. (2014). Enter the robot journalist: Users' perceptions of automated content. *Journalism Practice, 8*(5), 519–531.

Colestock, H. (2005). *Industrial robotics: Selection, design, and maintenance*. New York: McGraw-Hill.

Crews, J. (2016). *Robonomics: Prepare today for the jobless economy of tomorrow*. CreateSpace Independent Publishing Platform.

Crootof, R. (2015). War, responsibility, and killer robots. *North Carolina Journal of International Law and Commercial Regulation, 40*(4), 909–932.

Cubero, S. (Ed.). (2007). *Industrial robotics: Theory, modelling and control*. Mammendorf: Pro Literature Verlag Robert Mayer-Scholz.

Daugherty, P. R., & Wilson, H. J. (2018). *Human + machine: Reimagining work in the age of AI*. Boston, MA: Harvard Business Review Press.

DeCanio, S. J. (2016). Robots and humans—Complements or substitutes? *Journal of Macroeconomics, 49*, 280–291.

Drexler, K. E. (2013). *Radical abundance: How a revolution in nanotechnology will change civilization*. New York: Public Affairs.

Driessen, C., & Heutinck, L. F. M. (2015). Cows desiring to be milked? Milking robots and the co-evolution of ethics and technology on Dutch dairy farms. *Agriculture and Human Values, 32*(1), 3–20.

Dunis, C. L., Middleton, P. W., Karathanasopolous, A., & Theofilatos, K. A. (Eds.). (2017). *Artificial intelligence in financial markets: Cutting edge applications for risk management, portfolio optimization and economics*. London: Palgrave Macmillan.

Economic Times. (2018). Low skilled workers to face disadvantage as robotics, AI take over: ADB. *The Economic Times*, April 11. https://economictimes.indiatimes.com/jobs/low-skilled-workers-to-face-disadvantage-as-robotics-ai-take-over-adb/articleshow/63715526.cms

Ford, M. R. (2009). *The lights in the tunnel: Automation, accelerating technology and the economy of the future*. Acculant Publishing.

Ford, M. R. (2015). *Rise of the robots: Technology and the threat of a jobless future*. New York: Basic Books.

Frank, M., Roehring, P., & Pring, B. (2017). *What to do when machines do everything: How to get ahead in a world of AI, algorithms, bots and big data*. Hoboken, NJ: John Wiley & Sons, Inc.

Frey, C. B., & Osborne, M. A. (2017). The future of employment: How susceptible are jobs to computerisation? *Technological Forecasting and Social Change, 114*, 254–280.

Giarini, O., & Liedtke, P. M. (1996). *The employment dilemma: The future of work*. The Club of Rome.

Gould, S. J. (1996). *The mismeasure of man* (Revised & Expanded). W. W. Norton & Company.
Harteis, C. (Ed.). (2018). *The impact of digitalization in the workplace: An educational view* (Vol. 21). Cham: Springer.
Hill, J., Ford, W. R., & Farreras, I. G. (2015). Real conversations with artificial intelligence: A comparison between human–human online conversations and human–chatbot conversations. *Computers in Human Behavior, 49*, 245–250.
Huang, M. H., & Rust, R. T. (2018). Artificial intelligence in service. *Journal of Service Research, 21*(2), 155–172.
Ivanov, S. (2016). Will robots substitute teachers? *Yearbook of Varna University of Management, 9*, 42–47.
Ivanov, S. (2017). Robonomics—Principles, benefits, challenges, solutions. *Yearbook of Varna University of Management, 10*, 283–293.
Ivanov, S., & Webster, C. (2017). *The robot as a consumer: A research agenda.* Paper presented at the "Marketing: Experience and perspectives" Conference, 29–30 June 2017, University of Economics, Varna, Bulgaria, pp. 71–79.
Ivanov, S., & Webster, C. (2018). Adoption of robots, artificial intelligence and service automation by travel, tourism and hospitality companies—A cost-benefit analysis. In V. Marinov, M. Vodenska, M. Assenova, & E. Dogramadjieva (Eds.), *Traditions and innovations in contemporary tourism* (pp. 190–203). Newcastle upon Tyne: Cambridge Scholars Publishing.
Ivanov, S., Webster, C., & Berezina, K. (2017). Adoption of robots and service automation by tourism and hospitality companies. *Revista Turismo & Desenvolvimento, 27/28*, 1501–1517.
Kaplan, J. (2015). *Humans need not apply. A guide to wealth and work in the age of artificial intelligence.* Yale University Press.
Kauhanen, A. (2016). *The future of work: Challenges for men and women.* ETLA Brief No. 50. Retrieved from http://pub.etla.fi/ETLA-Muistio-Brief-50.pdf
Kaur, S. (2012). How medical robots are going to affect our lives. *IETE Technical Review, 29*(3), 184–187.
Kiggins, R. (2018). *The political economy of robots: Prospects for prosperity and peace in the automated 21st century.* Cham: Palgrave Macmillan.
Kurzweil, R. (2005). *The singularity is near. When humans transcend biology.* London: Duckworth Overlook.
LaGrandeur, K., & Hughes, J. J. (Eds.). (2017). *Surviving the machine age. Intelligent technology and the transformation of human work.* London: Palgrave Macmillan.
Lee, J. (2017). *Sex robots: The future of desire.* London: Palgrave Macmillan.
Lehman, T. (2015). Countering the modern Luddite impulse. *The Independent Review, 20*(2), 265–283.
Leonhard, G. (2016). *Technology vs. humanity.* Fast Future Publishing.
Lin, P., Abney, K., & Bekey, G. A. (2014). *Robot ethics: The ethical and social implications of robotics.* Cambridge, MA: The MIT Press.

Low, K.-H. (Ed.). (2007). *Industrial robotics: Programming, simulation and applications*. Mammendorf: Pro Literature Verlag Robert Mayer-Scholz.

Makridakis, S. (2017). The forthcoming Artificial Intelligence (AI) revolution: Its impact on society and firms. *Futures, 90*, 46–60.

Maurer, M., Gerdes, J. C., Lenz, B., & Winner, H. (Eds.). (2016). *Autonomous driving: Technical, legal and social aspects*. Berlin and Heidelberg: Springer Open.

McClure, P. K. (2017). "You're fired," says the robot: The rise of automation in the workplace, technophobes, and fears of unemployment. *Social Science Computer Review* (forthcoming). https://doi.org/10.1177/0894439317698637

Miller, M. R., & Miller, R. (2017). *Robots and robotics: Principles, systems, and industrial applications*. McGraw-Hill Education.

Min, H. (2010). Artificial intelligence in supply chain management: Theory and applications. *International Journal of Logistics Research and Applications, 13*(1), 13–39.

Mirheydar, H. S., & Parsons, J. K. (2013). Diffusion of robotics into clinical practice in the United States: Process, patient safety, learning curves, and the public health. *World Journal of Urology, 31*(3), 455–461.

Murphy, J., Hofacker, C., & Gretzel, U. (2017). Dawning of the age of robots in hospitality and tourism: Challenges for teaching and research. *European Journal of Tourism Research, 15*, 104–111.

Neapolitan, R. E., & Jiang, X. (2013). *Contemporary artificial intelligence*. Boca Raton, FL: CRC Press.

Nørskov, M. (2016). *Social robots: Boundaries, potential, challenges*. London and New York: Routledge.

NPR. (2011). Science diction: The origin of the word 'robot'. Retrieved from https://www.npr.org/2011/04/22/135634400/science-diction-the-origin-of-the-word-robot

NPR. (2015). Map: The most common job in every state. Retrieved from https://www.npr.org/sections/money/2015/02/05/382664837/map-the-most-common-job-in-every-state

Organisation for Economic Co-operation and Development/OECD/. (2017). *The next production revolution: Implications for governments and business*. Paris: OECD Publishing.

Pires, J. N. (2007). *Industrial robots programming: Building applications for the factories of the future*. New York: Springer US.

Preparing for The Future of Artificial Intelligence. (2016). Published by the Executive Office of the President of the USA, National Science and Technology Council and the Committee on Technology, 2nd October 2016. Retrieved October 14, 2016, from https://www.whitehouse.gov/sites/default/files/whitehouse_files/microsites/ostp/NSTC/preparing_for_the_future_of_ai.pdf

Remus, D. & Levy, F. (2015) Can robots be lawyers? Computers, lawyers, and the practice of law. *SSRN Working paper*. Retrieved from http://ssrn.com/abstract=2701092

Royakkers, L., & van Est, R. (2016). *Just ordinary robots: Automation from love to war*. Boca Raton: CRC Press.
Russell, S. J., & Norvig, P. (2016). *Artificial intelligence: A modern approach*. Harlow: Pearson Education Limited.
Santens, S. (2017). Unconditional basic income as a solution to technological unemployment. In K. LaGrandeur & J. J. Hughes (Eds.), *Surviving the machine age. Intelligent technology and the transformation of human work* (pp. 107–116). London: Palgrave Macmillan.
Schommer, E., Patel, V. R., Mouraviev, V., Thomas, C., & Thiel, D. D. (2017). Diffusion of robotic technology into urologic practice has led to improved resident physician robotic skills. *Journal of Surgical Education, 74*(1), 55–60.
Schwab, K. (2016). *The fourth industrial revolution*. Cologny and Geneva: World Economic Forum.
Sendler, U. (Ed.). (2018). *The internet of things: Industrie 4.0 unleashed*. Berlin: Springer Vieweg.
Shanahan, M. (2015). *The technological singularity*. Cambridge, MA: The MIT Press.
Sheahen, A. (2012). *Basic income guarantee: Your right to economic security*. New York: Palgrave Macmillan.
Skilton, M., & Hovsepian, F. (2018). *The 4th industrial revolution: Responding to the impact of artificial intelligence on business*. Cham: Palgrave Macmillan.
Solon, O. (2016, June 17). Self-driving trucks: What's the future for America's 3.5 million truckers? *The Guardian*. Retrieved from https://www.theguardian.com/technology/2016/jun/17/self-driving-trucks-impact-on-drivers-jobs-us
Sparrow, R. (2007). Killer robots. *Journal of Applied Philosophy, 24*(1), 62–77.
Stern, A. (2016). *Raising the floor: How a universal basic income can renew our economy and rebuild the American dream*. Public Affairs.
Stevens, Y. A., & Marchant, G. E. (2017). Policy solutions to technological unemployment. In K. LaGrandeur & J. J. Hughes (Eds.), *Surviving the machine age. Intelligent technology and the transformation of human work* (pp. 117–130). London: Palgrave Macmillan.
Swan, M. (2017). Is technological unemployment real? An assessment and a plea for abundance economics. In K. LaGrandeur & J. J. Hughes (Eds.), *Surviving the machine age. Intelligent technology and the transformation of human work* (pp. 19–33). London: Palgrave Macmillan.
Talwar, R. (Ed.). (2015). *The future of business*. Fast Future Publishing.
Talwar, R., Wells, S., Whittington, A., Koury, A., & Romero, M. (2017). *The future reinvented. Reimagining life, society, and business*. Fast Future Publishing.
The National Artificial Intelligence Research and Development Strategic Plan. (2016). Published by the Executive Office of the President of the USA, National Science and Technology Council and the Committee on Technology,

2nd October 2016. Retrieved October 14, 2016, from https://www.whitehouse.gov/sites/default/files/whitehouse_files/microsites/ostp/NSTC/national_ai_rd_strategic_plan.pdf

Timms, M. J. (2016). Letting artificial intelligence in education out of the box: Educational cobots and smart classrooms. *International Journal of Artificial Intelligence in Education, 26*(2), 701–712.

Warwick, K. (2012). *Artificial intelligence: The basics.* Oxon: Routledge.

Watson, P. (2018, August 23). There's only one way to solve our stagnant growth problem. *Forbes.* Retrieved from https://www.forbes.com/sites/patrickwwatson/2018/08/23/theres-only-one-way-to-solve-our-stagnant-growth-problem/#60cb57462e1d

West, D. M. (2018). *The future of work: Robots, AI, and automation.* Washington: Brookings Institution Press.

Wiener, A. (2017, November 29). Inside Adidas' robot-powered, on-demand sneaker factory. *Wired.* https://www.wired.com/story/inside-speedfactory-adidas-robot-powered-sneaker-factory/

Wirtz, J., Patterson, P., Kunz, W., Gruber, T., Lu, V. N., Paluch, S., & Martins, A. (2018). Brave new world: Service robots in the frontline. *Journal of Service Management, 29*(5), 907–931. in press.

World Economic Forum. (2016). *The future of jobs. Employment, skills and workforce strategy for the fourth industrial revolution.* World Economic Forum report. Retrieved from http://www3.weforum.org/docs/WEF_Future_of_Jobs.pdf

World Economic Forum. (2018). *The future of jobs report 2018.* World Economic Forum report. Retrieved from http://www3.weforum.org/docs/WEF_Future_of_Jobs_2018.pdf

Xu, A., Liu, Z., Guo, Y., Sinha, V., & Akkiraju, R. (2017). A new chatbot for customer service on social media. In *Proceedings of the 2017 CHI Conference on Human Factors in Computing Systems* (pp. 3506–3510). New York: ACM.

CHAPTER 9

The Dark Side of Big Data: Personal Privacy, Data Security, and Price Discrimination

Yang Liu and Connor Greene

Introduction

According to Furman and Simcoe (2015), the Council of Economic Advisers released a study, "Big Data and Differential Pricing", in February 2015. In terms of this study (2015), big data is defined as the capability to collect a huge amount of data from various sources for investigating, analyzing, and understanding consumers, and forecasting market demand and affordable prices for consumers. In turn, the way firms do business has been changed by this new technology. Consequently, the whole of society has been changed, with this new technology both bringing benefits and generating challenges.

Y. Liu (✉)
Fitchburg State University, Fitchburg, MA, USA
e-mail: yliu4@fitchburgstate.edu

C. Greene
Southern New Hampshire University, Hooksett, NH, USA
e-mail: connor.greene@snhu.edu

B. George, J. Paul (eds.), *Digital Transformation in Business and Society*, https://doi.org/10.1007/978-3-030-08277-2_9

Benefits and Challenges to Business and Society since the Big Data Era

Personal Privacy and Data Security

When the big data era finally arrives, firms will cheer up because of this smart technology for which they have been waiting for years. Web browsers or the owners of websites normally generate a "cookie", which is a file on a user's computer. This file saves all the information on the user. Therefore, web browsers or the owners of websites can draw a picture from the details of any user based on the information they accessed. TRUSTe, a technology compliance and security company, reported in 2014 that the top 100 websites browsed the most are monitored by more than 1300 firms. Some of the data collected by these firms is "shared" (also known as "piggybacking") with outsiders. Normally this monitoring process is unknown to web users. For example, an advertising technology company, RadiumOne, puts cookies on users without notifying them, when they click on a link sent by a friend (Big Data and Differential Pricing, 2015; Chen, Choe, & Matsushima, 2018; The Economist, 2014).

New information technology platforms offer applications (or apps) for maps, search engines, social media, and so on to enrich people's daily lives. In turn, the omnipresent online search, payment transactions, and social interactions leave a trail of personal information, containing emotions, tastes, behaviors, locations, affordable prices, and so on. These apps do not support cookies, but they can access and record the unique device identification (UDID) number of the mobile device to recognize individual users. PubMatic is a company helping publishers increase sales. It provides around 50 to 70 points of user information on desktops and around 100 on mobiles, which includes the mobile device user's location (Chen et al., 2018; The Economist, 2014).

New Business Pattern

According to Kamenica, Mullainathan, and Thaler (2011), the more information cell-phone providers and data collectors gather and analyze, the more they understand consumer behavior. Given enough time to collect enough data, data analytics can understand a consumer more than him- or herself. Calo (2014) explains personalized advertising as the ability of firms to detect when is the best time to deliver a commercial or how to build an emotional relationship with consumers based on their personal experience. Duhigg (2012) observes that Target can distinguish pregnant consumers based on their shopping record and directly deliver related advertisements regarding baby products to them.

These big data enable firms to reach their target consumers directly and save millions of dollars that they are wasting on people who will not purchase their goods or services no matter how hard they try, because the firms are able to analyze consumers' current situation, such as pregnancy, divorce, relocation, and so on. Before the big data era, marketing researchers would design, deliver, collect, and analyze surveys to understand their potential consumers. This process took months, even years, to provide a rough understanding based on a sample group that was never big enough. All business plans and marketing designs would be built on that information, until the day that big data ended it.

Big data change the way firms do business, make marketing plans, and deliver advertisements. With big data, firms can directly locate a potential consumer, deliver an advertisement at an appropriate time, and charge an affordable price to make a deal. For example, when consumers search for a business suit online, advertisements do not pop up randomly but in a logical layout regarding what they purchased before. This technology matches consumers with the goods or services they are looking for. Consumers can benefit from this personalized advertising as well, since it saves them time. They can find what they need more quickly.

Price Discrimination

On the other hand, big data bring new challenges to this world as well. There are some side effects accompanying big data. Nowadays we face new situations that we could not imagine years ago, such as personal privacy, data security, and price discrimination.

Big Data and Differential Pricing (2015) points out that one main concern with big data is price discrimination, which means charging different consumers different prices for similar products. Price discrimination can be divided into three types. First-degree price discrimination is personalized pricing, meaning there is no fixed sales price. Every consumer can negotiate the price, and they eventually pay different prices. Second-degree price discrimination is quantity discounts. In other words, buy more, save more. Third-degree price discrimination is a demographic discount, which means a specific group of people can get specific treatment. For example, universities provide students who serve in the military with a specific scholarship or free tuition (Big Data and Differential Pricing, 2015).

In some cases, even though price discrimination sounds unfair, it is not, due to the factual situation. For example, children normally can get a specific discount, not only because they are in a specific age range but also because they increase the expenditure with their entire family. For example, parents have to accompany their children to the Disneyland, the cinema, the arcade, and so on. Therefore, children stimulate demand for certain goods and services that would not otherwise be consumed by adults. An airplane ticket is more expensive when the booking day is closer to the departure day, due to the increase in overall demand on that given date, as well as the finite number of available seats.

In some other cases, price discrimination may be really unfair. According to Townley, Morrison, and Yeung (2017), Amazon.com sold DVDs to different people for different prices in 2000. After a public protest, Amazon refunded the consumers who were charged a higher price and claimed that this was a test. However, a similar issue happened again in 2012. A consumer found that the price of a set of mahjong tiles in her shopping basket suddenly jumped from $54.99 to $79.99; when she emptied the basket and added the same goods into it again, the price went back to $59.99. The *Wall Street Journal* reported in 2012 that Staples Inc. was displaying different prices to people based on their location. Staples.com showed a lower price to consumers who lived within 20 miles of a rival store (Townley et al., 2017). According to Michael and Mohammed (2015), in international travel bookings for US hotel and car rental, the situation is the same.

According to Newcomer (2017), Uber Technologies Inc. uses a "route-based pricing" system, which charges consumers differently depending on the expectation of how much they would like to pay. This system is different from the traditional pricing model, which charges consumers only based on the quantity of a service they purchase. The pricing system Uber used before calculates fares by adding up time, mileage, and multipliers based on geographic demand. The new "route-based pricing" system charges some consumers a higher price not because they ride to a farther location, but because it is considered they would pay a higher price.

A similar situation was reported by Consumer Intelligence Research Partners (CIRP) in 2013: that among all smartphone consumers, iPhone consumers were charged the highest average monthly carrier bill in the

USA. CIRP co-founder Michael Levin believes this situation happens not because consumer behaviors are different among smartphone users, but because of their data plans and carriers (Keller, 2013; Paczkowski, 2013). In other words, iPhone users are charged a higher price for their carrier bills in the USA, not because the same apps cost more in data streaming than on an Android phone, nor because iPhone users spend longer on their phone than Android users, but possibly because they are considered richer, since the price of an iPhone is higher than that of an Android phone.

More evidence can be observed internationally. For example, according to Wu (2018), among 2008 participants, 51.3% suffered price discrimination when they purchased tickets, booked hotels, and subscribed for memberships online in China. There are several different types of price discrimination. (1) More refreshes, higher price. When booking flight tickets and hotels, the more times consumers refresh the website, the higher the price they will see. Once consumers clean the website's cache, cookies, and history, the price decreases. (2) Members pay more. Consumers may purchase a membership for movie ticket discounting, but some of them find that their price is higher than for consumers who do not have a membership. (3) Consumers pay a higher price on iPhone than on Android. Many iPhone consumers are charged a higher price when they subscribe to a membership or pay a taxi. Some firms explain this situation by saying that mobile payment services (Apple Pay, Alipay, etc.) offer convenience to their clients and users while charging a higher service fee on the iPhone than on Android.

Machlup (1955) may not be the first economist who has noticed that the definition of price discrimination ("the practice of a firm selling a homogeneous commodity at the same time to different purchasers at different prices") is somewhat vague, but he is the first to clarify this definition as "the practice of a firm or group of firms of selling (leasing) at prices disproportionate to the marginal costs of the products sold (leased) or of buying (hiring) at prices disproportionate to the marginal productivities of the factors bought (hired)". Moreover, according to Thomas (2012), in the insurance context the concept of price discrimination, which is selling products with similar marginal costs at different prices to different clients, is "non-risk" price discrimination. However, insurance is a risk-related product. Various price segments of it exhibit different marginal costs. This is why the data on (potential) consumers' health is so important for both insurance companies and consumers. With the data, the insurance company can charge a consumer a personalized price which may be significantly different from the price without the data. Consumers may even do not know that their price is personalized and higher than that for others. Big Data and Differential Pricing (2015) explains the risk-based pricing policy of insurance companies as finally helping the whole of society by charging drivers with risky behaviors or a history of traffic accidents a higher price. Therefore, these drivers will change their behaviors and society will become safer for everyone.

It is not hard to understand that firms have plenty of reasons to treat their clients differently. This is considered necessary and motivation for them to pay for big data, especially service firms. Different from goods companies, service companies have a long-term relationship with their clients. They provide services constantly, which means the relationship between a service company and a consumer keeps changing with time. For instance, insurance companies service their clients from cradle to grave. Some clients have good health while some others do not. Therefore, insurance companies could separate them into different segments (high risk, moderate, low risk) and foresee the future revenue compared to the expected underwriting expense. Thus, insurance companies charge different prices to match the health situation of their clients.

Discussion and Conclusion

Podesta, Pritzker, Moniz, Holdren, and Zients (2014) believe that it deserves more serious concern when data impact decisions on individuals' credit, housing, education, healthcare, or employment. Moreover, personal privacy is another critical concern (Big Data and Differential Pricing, 2015). According to Payton, Schmidt, and Claypoole (2014), the core of privacy is not about information but about choice. According to Hacker and Petkova (2017), both greater social equality and inequality have been generated with big data. Because of big data, technology, education, and business are more efficient than before. However, consumer protection and discrimination bring greater inequality to society. Big data provide firms with opportunities for personalizing services. Before the new big data era, consumers were exposed to the same advertisements. They had an equal chance with the same offers at the same price. However, big data collection and analytics are changing this situation. With the new technology, firms know exactly what consumers want and how much they can afford. On the other hand, consumers do not know that they are being treated differently from one another. In other words, with detailed data, companies can manipulate their consumers individually. Consumers are guided by an invisible hand to a place they think will satisfy their needs. However, they may pay a higher price or make a suboptimal decision because of firms obtaining their data and using them inappropriately. Big Data and Differential Pricing (2015) suggests that enhancing the transparency of how firms collect, use, and trade consumers' data would promote competition in the market and provide consumers with choices. In addition, Hacker and Petkova (2017) point out that transparency is important during this new big data era. Citron and Pasquale (2014) state that people have the right to access and correct their information and they have the right to know how they are rated and ranked as well. In other words, consumers have the right to keep their privacy, especially when they do not buy insurance from certain insurance companies. In addition, even though it may be necessary for insurance companies to obtain these data to lower their risks and costs, do they have the right to do so?

First of all, who owns the data? Or we may consider this from another angle: where is the line drawn when collecting data without notifying people that their data are collected, retained, and will be used to manipulate them later? Before the big data era, when companies had to collect data by sending out surveys, they had to send an obvious notification to the participants regarding this concern. They had to tell the participants it was voluntary. In some cases, they paid participants to collect their data. Moreover, normally, the participants were anonymous. They did not provide their personal information, such as name, date of birth, and address. Therefore, they did not need to worry about how these data would be used. However, the new technology changed this. The new technology enables web browsers and apps to collect and retain personal data on a continual basis. Because of this process, eventually, consumers get personalized advertisements. In some cases, this is harmless. On the contrary, it is convenient. On the other hand, it is a trap. Under the manipulation of a marketing strategy provided to companies based on specific data analytics, consumers purchase something that they do not need or are charged a higher price just because they are (considered) richer than others.

Consumers may not know that their data are being collected and their behaviors are being manipulated. They may enjoy the convenience. For example, a pregnant woman does need to prepare for the upcoming baby, and advertisements and coupons for baby products are useful to her. Or for a resident who has just moved to a new apartment, information about furniture discounts and internet services is helpful.

However, price discrimination is not a convenient experience. For example, ticket selling or hotel booking websites manipulate consumers by increasing prices according to their refresh frequency. This manipulates consumers to close the deal as soon as possible by implying that the price will keep increasing or that tickets are selling out. Consumers waste money as a result of their actions being recorded and analyzed by data analytics.

In conclusion, we believe that transparency of data collection is important and necessary. Individuals have the right to access data held on them and to have greater control over their own information. They should be notified when their information is collected and traded. They should have the right to say no to certain firms whose business is to dig, gather, collect, and analyze data. Furthermore, firms should be restricted by regulations regarding their operations. On the other hand, consumers who would like to enjoy personalized services, such as advertisements or coupons, can choose to allow data-gathering firms to collect their data with their permission.

References

Big Data and Differential Pricing. (2015). *The White House President Barack Obama*. Retrieved from https://obamawhitehouse.archives.gov/sites/default/files/whitehouse_files/docs/Big_Data_Report_Nonembargo_v2.pdf

Calo, R. (2014). Digital market manipulation. *George Washington Law Review, 82*(4), 995–1051. 57p. Database: Business Source Ultimate.

Chen, Z., Choe, C., & Matsushima, N. (2018). Competitive personalized pricing. *ISER Discussion Paper*. Retrieved from https://www.monash.edu/business/economics/research/publications/publications2/0218Copetitive chenchoe.pdf

Citron, D. K., & Pasquale, F. A. (2014). The scored society: Due process for automated predictions. *Washington Law Review, 89*(1), 33. Washington Law Review Association.

Duhigg, C. (2012). How companies learn your secrets. *New York Times Magazine*, February 16. Retrieved from https://www.nytimes.com/2012/02/19/magazine/shopping-habits.html

Furman, J., & Simcoe, T. (2015). The economics of big data and differential pricing. *The White House President Barack Obama*. Retrieved from https://obamawhitehouse.archives.gov/blog/2015/02/06/economics-big-data-and-differential-pricing

Hacker, P., & Petkova, B. (2017). Reining in the big promise of big data: Transparency, inequality, and new regulatory frontiers. *Northwestern Journal of Technology & Intellectual Property, 15*(1), 1–42. Northwestern University.

Kamenica, E., Mullainathan, S., & Thaler, R. (2011). Helping consumers know themselves. *The American Economic Review, 101*(3), 417–422. American Economic Association.

Keller, J. (2013). iPhone users pay higher cell phone bills than any other smartphone user. *iMore*. Retrieved from https://www.imore.com/iphone-users-biggest-cash-cows-carriers

Machlup, F. (1955). Characteristics and types of price discrimination. NBER chapters. In *Business concentration and price policy* (pp. 397–440). National Bureau of Economic Research, Inc.

Michael, R., & Mohammed, R. (2015). Who's paying more to tour these United States? Price differences in international travel bookings. *Technology Science*.

Newcomer, E. (2017, May 19). Uber starts charging what it thinks you're willing to pay. *Bloomberg*. Retrieved from https://www.bloomberg.com/news/articles/2017-05-19/uber-s-future-may-rely-on-predicting-how-much-you-re-willing-to-pay

Paczkowski, J. (2013). iPhone users rack up the highest carrier bills. *All Things D*. Retrieved from http://allthingsd.com/20130130/wireless_bills_by_os_android_ios/

Payton, T., Schmidt, H. A., & Claypoole, T. (2014). *Privacy in the age of big data: Recognizing threats, defending your rights, and protecting your family*. Lanham, MD: Rowman & Littlefield Publishers.

Podesta, J., Pritzker, P., Moniz, E. J., Holdren, J., & Zients, J. (2014). Big data: Seizing opportunities, preserving values. *Executive Office of the President*. Retrieved from https://obamawhitehouse.archives.gov/sites/default/files/docs/20150204_Big_Data_Seizing_Opportunities_Preserving_Values_Memo.pdf

The Economist. (2014, September 11). Getting to know you. Retrieved from https://www.economist.com/special-report/2014/09/11/getting-to-know-you

Thomas, R. (2012). Non-risk price discrimination in insurance: Market outcomes and public policy. *The Geneva Papers on Risk and Insurance. Issues and Practice, 37*(1), 27–46. Palgrave Macmillan.

Townley, C., Morrison, E., & Yeung, K. (2017). Big data and personalized price discrimination in EU competition law. *Yearbook of European Law, 36*(1), 683–748. 66p. Publisher: Oxford University Press/USA.

Wu, Y. Q. (2018). 用苹果手机打车、看电影、买机票比安卓贵?记者亲测被大数据"杀熟". *Shen Zhen Shang Bao*. Retrieved from http://www.sohu.com/a/226569266_355791

CHAPTER 10

Digital Finance for Financial Inclusion and Inclusive Growth

Md. Nur Alam Siddik and Sajal Kabiraj

Introduction

The notion of financial inclusion emerged in early 2000 and thus was given policy priority by governments and international organizations to reduce poverty and achieve inclusive growth. Consequently, governments and policy makers formulated policies to foster financial inclusion in developing and emerging economies (GPFI, 2010). Nowadays, digital finance has gained much attention from policy makers as a mechanism for reducing poverty and achieving financial stability, because digital finance expands financial inclusion, which ultimately promotes financial stability.

Digital finance has numerous benefits to all stakeholders, including financial service users, digital finance service providers, governments, policy makers and the economy. For example, it expands access to finance among the poor and disadvantaged, it can reduce the cost of service for

Md. N. A. Siddik (✉)
Begum Rokeya University, Rangpur, Bangladesh
e-mail: nasiddik@brur.ac.bd

S. Kabiraj
Dongbei University of Finance and Economics, Dalian, China
e-mail: skabiraj@dufe.edu.cn

B. George, J. Paul (eds.), *Digital Transformation in Business and Society*, https://doi.org/10.1007/978-3-030-08277-2_10

banks and fintech suppliers by employing a branchless technology-based banking system, and it can lead to a rise in total expenditure for governments. The World Bank Group President Jim Yong Kim indicated that by 2020, digital financial instruments such as e-money cards, debit cards and low-cost bank accounts will significantly expand the scope of financial access for excluded people.[1] Notwithstanding its benefits, the issue of digital finance, financial inclusion and financial stability has not been addressed in the literature. This study aims to fill this gap by addressing the link among digital finance, financial inclusion and financial stability.

This research contributes to the financial inclusion campaigns initiated by the World Bank as an operational solution for poverty eradication in developing and emerging economies. The findings of this study can help policy makers to understand the issues related to the prompt expansion of digital financial services, strategies for its effective delivery to the poor and the risks involved in digital financial inclusion. Secondly, this study adds to the existing literatures on emerging digital finance and financial inclusion, which will enable academics and researchers to attend to the effects of digital financial services on financial inclusion.

Conceptualizing Digital Finance

Manyika, Lund, Singer, White, and Berry (2016) defined digital finance as financial services delivered through a digital infrastructure such as a mobile phone and internet network, which eventually encourage less usage of cash and the traditional brick-and-mortar banking system.

According to Gomber, Koch, and Siering (2017), digital finance comprises a magnitude of new financial products, financial services, financial management software and new ways to communicate and interact with customers. Digital finance implies a procedure of a certain blend of related money management tools that are conveyed using a portable or Web-based system (Peake, 2012).

Klapper, El-Zoghbi, and Hess (2016) demonstrate that digital finance permits people to access finance from distant places and at any time, which enables them to manage their funds well at times of need and thus shrinks the possibility that they will fall into poverty.

While there is no universal description of digital finance, there is a consensus that it incorporates financial products, financial services, technolo-

[1] https://www.worldbank.org/en/news/speech/2016/04/14/remarks-by-world-bank-group-president-jim-yong-kim-at-the-global-connect-initiative

gies and infrastructure that empower people to have access to payments, savings and credit facilities along with other required financial services through online means and without a physical visit to a bank branch.

Conceptualizing Financial Inclusion

Financial inclusion aims to ensure access to and use of financial services such as accounts, savings and credit for everyone, including the poor and disadvantaged. It targets the unbanked, people who have not been offered financial services, bringing them under the umbrella of the formal financial system, and the under-banked, people who have access to but have not used financial services. The overriding goal is to create an environment in which all people will have easy access to financial services to ensure the participation of everyone in the course of economic growth and receiving the benefits of economic growth accordingly.

Siddik, Sun, and Kabiraj (2015) describe financial inclusion as a means of providing easier and timely access to finance to all sections of society, including disadvantaged groups, which in turn will promote inclusive growth.

Hannig and Jansen (2010) state that financial inclusion programs aim to offer greater opportunity with easier access to proper financial services to people, including disadvantaged groups, which in turn fosters the economic development of a country. Sarma (2015) defines financial inclusion as a procedure which warrants easier access, readiness and usage of financial services for all people of a country. This explanation strictly emphasizes access to and usage of formal financial services for entire segments of society, which is one of the most vital factors in the economic growth of a nation. Following the definition of Sarma (2015), we focus on the access and usage dimensions of financial inclusion and define it as a process which aims to expand the financial network among people including the unbanked and the under-banked. In developing economies, the banking sector is reflected as the key pillar and the most contributive sector to economic growth. Thus, we focus on financial inclusion by commercial banks.

Effect of Digital Finance on Financial Inclusion

The impact of digital finance on financial inclusion is now evident. Digital financial services have the substantial prospect of magnifying the delivery of financial services to a large segment of society, including the poor, through innovative technologies such as mobile banking solutions, e-money systems and digital payment infrastructure at a reasonable cost, in appropriate modes and in a safe environment. According to the 2016 annual report of the International Telecommunication Union (ITU), 95% of the world population lives in areas which are covered by a mobile cellular signal and in 2016, 47% of the world population was online, which indicates progress, but at the same time over half of the world's population, or roughly 3.9 billion people, was not online. This suggests that there is huge potential to settle financial matters digitally.

Digital finance adds new stratums to financial inclusion, proposing new methods of expanding the inclusion of groups profiled as poor and disadvantaged in mainstream financial activities (Gabor & Brooks, 2017).

Manyika et al. (2016) point out that the wide-ranging usage of digital finance could increase the annual gross domestic product (GDP) of all economies by 6% by 2025. Arguing the same, Ghosh (2016) employs data from 2001–2012 on Indian states and observes that mobile telephony as a means of digital financial services has significant effects on economic growth. Proposing similar contributive effects of digital finance on financial inclusion, we first examine world data and then provide case studies on India.

Methodology

To determine whether digital finance promotes financial inclusion, in this section we present the effects of digital finance on financial inclusion. In this study financial inclusion is measured by outstanding loans with commercial banks (% of GDP; log of outstanding loans or LOL) and digital finance is measured by three widely used proxies: the number of automated teller machines (ATMs) per 100,000 adults (ATM); the percentage of the population using the internet (INT); and mobile money transaction value (percentage of GDP; MMT). Based on the availability of data, we collected data from 2004 to 2016 on 189 countries. Variable measurements and data sources are presented in Table 10.1. Taking the average of each variable for each year, firstly we conducted trend analysis and secondly we produced two case studies in support of our hypothesis.

Empirical Findings

We divide this section into two parts. In the first part we conduct a trend analysis with linear regression, and in the second part we provide case studies. In Figs. 10.1, 10.2 and 10.3 we represent the financial inclusion

Table 10.1 Variables, their proxies, legend and source

Variable	*Measurement*	*Legend*	*Sources*
Financial inclusion	Natural log of outstanding loans with commercial banks (% of GDP)	LOL	FAS*
Digital finance	Natural log of number of automated teller machines (ATMs) per 100,000 adults	ATM	FAS
	Individuals using the internet (% of population)	INT	WDI**
	Mobile money transactions: value (% of GDP)	MMT	FAS

* Financial Access Survey
** World Development Indicator

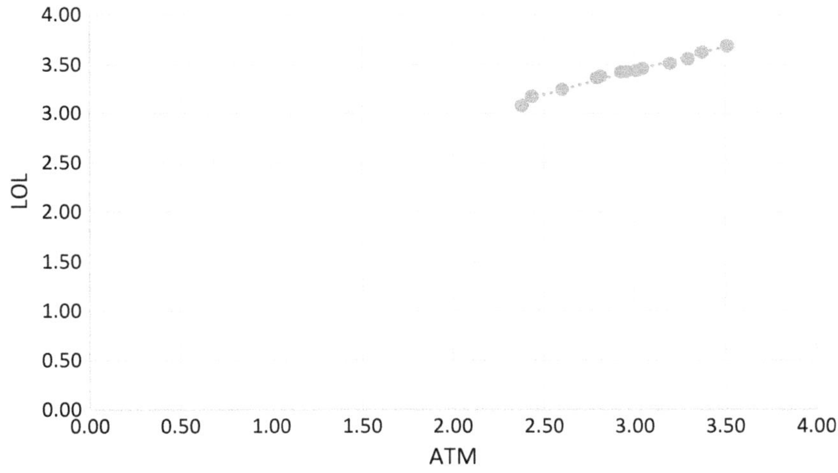

Fig. 10.1 Impact of ATM on LOL

variable, LOL, on the Y axis and digital finance proxies on the X axis. These figures represents the association of digital finance proxies (ATM, INT and MMT) with financial inclusion, LOL. Figure 10.1 depicts an upward, straight trend line between the number of ATMs per 100,000 adults and outstanding loans with commercial banks, which means that there is a positive relation between ATM and LOL.

Figure 10.2 exhibits the relation between internet penetration and financial inclusion. With linear trend analysis, we see that the higher the number of internet users, the higher the degree of financial inclusion. In other words, there is a positive effect of INT on LOL.

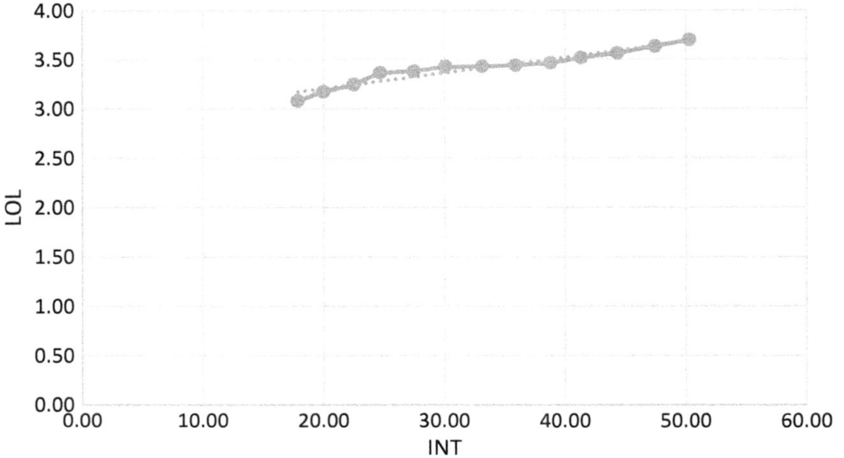

Fig. 10.2 Impact of INT on LOL

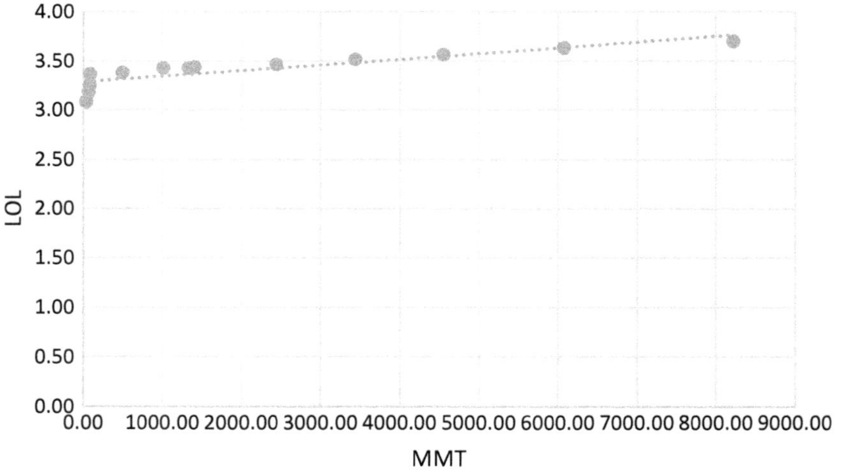

Fig. 10.3 Impact of MMT on LOL

The third proxy we use is the value of mobile money transactions. Since mobile money is one of the most widely used instruments of digital finance which has great global penetration, we have used this proxy to examine whether this contributes to financial inclusion or not. The results of the trend analysis indicate that there is positive movement between MMT and LOL. Thus, we argue that digital finance promotes financial inclusion. Our findings are similar to those of Ozili (2018).

Case Study

In this section we look at India, which is well and truly in the digital era and is a good example of digital financial inclusion. The essence of the case study is the use of technology to support the country's flagship benefit transfer programs and open doors for the poor to the financial mainstream. India is a leading case study of commitment to digital transformation and financial inclusion. With over a billion mobiles, 325 million broadband connections and 306 million new bank accounts, the country has become a case study in digital financial inclusion. The 500-million mobile connections in rural areas offer an opportunity to achieve full financial inclusion.

To make this happen, India has been launching strategic innovations to improve the design and drive a faster rollout of large benefit programs. The innovations pointedly track the delivery of entitlements to beneficiaries and help engage with the excluded millions using mobile-based tools. India's Digital Financial Inclusion program has piloted an integrated cashless digital platform to bring the farming community to center stage, while supporting a technological solution to fund and track rural projects. Another pilot infuses transparency into the delivery of nutrition across anganwadis, or rural childcare centers. A notable new entry is a technology platform to support a nationwide welfare scheme for pregnant and lactating mothers.

The Digital Financial Inclusion program uses insights drawn from work in both the corporate social responsibility (CSR) and government spaces to develop collaborative information technology (IT) platforms that bring corporates and governments together for community development. Moving upstream to benefit design, the program has also deployed analytical tools that use complex information on existing government schemes and databases to help design new programs.

Together with non-governmental organizations (NGOs) and the government, India's Digital Financial Inclusion program collaborates, conceptualizes and develops innovative solutions for enabling direct benefit transfer (DBT), financial inclusion and financial literacy. The program conducts research studies to understand and analyze factors that could possibly lead to the greater adoption and use of digital services by the poor. Postal accounts offering digital bank accounts could lead the way in this. Recently, the India Post Payments Bank (IPPB) has offered facilities for deposits, money transfers, DBTs and bills and utility payments, among other things.

Case 1: India Post Payments Bank: A Case Study of Digital Financial Inclusion

IPPB offers three types of zero-balance savings accounts: regular savings, digital savings and basic savings bank deposit accounts.

October 9 is celebrated as World Post Day. It also marks the beginning of National Postal Week, which is celebrated until October 15. India Post, which runs postal services in the country, sent greetings to its followers on the occasion, as did the government to its employees. India Post, which has a network of 155,015 post offices, also started the India Post Payments Bank in the year of 2018. IPPB has 650 branches and 3250 access points.

IPPB offers three types of savings account: regular savings account, digital savings account and basic savings bank deposit account.

- *India Post Payments Bank regular savings account*

This regular savings account allows for unlimited cash deposits and withdrawals. The zero balance account offers an interest rate of 4% per annum on end-of-day balances. The interest is paid on a quarterly basis.

This account does not require any monthly average balance and it can be opened with a zero balance. The maximum end-of-day balance that can be maintained is USD 1458.61.

- *India Post Payments Bank digital savings account*

IPPB's mobile app allows customers to open this account on their own, instantly. A maximum end-of-day balance of USD 1458.61 is allowed and a maximum yearly cumulative deposit of USD 2917.73.

This zero balance account is subject to closure if "know your customer" (KYC) verification is not completed within 12 months of opening. KYC helps banks verify their customers' identities. Aadhaar and PAN cards are mandatory for opening a digital savings account at IPPB.[2]

- *India Post Payments Bank basic savings bank deposit account*

This savings account has all the features and benefits offered by the regular savings account, except that it allows only four cash withdrawals in a month. It also does not require maintenance of any monthly average balance. It can be opened with a zero balance and offers an interest rate of 4% annum, paid quarterly.[3]

Financial inclusion products must be founded on the following principles:

E1: Effectiveness
How effective it is to use, how useful the product is and how it can make a difference in daily life.
E2: Efficiency
How efficient it is to use, whether it is easy to use, easy to handle, child compliant, suitable for senior citizens and differently abled people.
E3: Economical
How affordable it is for people at the bottom of the pyramid. The pricing of the product should be according to the cost of the technology. It should be accessible in small quantities to the majority of consumers.
E4: Environment
The product should be sustainable, eco-friendly and able to be recycled easily.

The earnings from these products can be reflected as a measure of the 3Ps (People, Planet and Profits):

$$\text{Earnings} = E1, E2, E3, E4 \propto P1, P2, P3$$

[2]Aadhaar number is a unique 12-digit biometric-based random number issued by the UIDAI (Unique Identification Authority of India) to the residents of India after satisfying the verification process and the enrolment process is totally free of cost. A permanent account number (PAN) is a unique 10-digit alphanumeric number, issued by the Indian Income Tax Department, to any "person" who applies for it or to whom the department allots the number without an application and it is valid across India and is unaffected by a change of name, address within or across states in India or other factors.

[3]Sunil Dhawan (2017), 'This savings account doesn't require any minimum balance: Find out if it suits you', article published in Economic Times, India, dated 29 September 2017.

Digital finance products could be subjected to A/B testing. In web analytics, A/B testing (bucket tests or split-run testing) is a randomized experiment with two variants, A and B. It includes the application of statistical hypothesis testing or "two-sample hypothesis testing" as used in the field of statistics. A/B testing is a way to compare two versions of a single variable, typically by testing a subject's response to variant A against variant B, and determining which of the two variants is more effective.

The two versions that are compared are identical except for one variation that might affect a user's behavior. Version A might be the currently used version (control), while version B is modified in some respect (treatment). For instance, on an e-commerce website the purchase funnel is typically a good candidate for A/B testing, as even marginal improvements in drop-off rates can represent a significant gain in sales.

A/B testing has been marketed by some providers as a change in philosophy and business strategy in certain niches, though the approach is identical to a between-subjects design, which is commonly used in a variety of research traditions. A/B testing as a philosophy of web development brings the field into line with a broader movement toward evidence-based practice. The benefits of A/B testing are considered to be that it can be performed continuously on almost anything, especially since most marketing automation software now typically comes with the ability to run A/B tests on an ongoing basis.

Based on the results of A/B testing, products could be redesigned to meet the specific needs of customers within a period of 24 hours. Real-time tracking may not be possible due to the small amount of capital involved. However, A/B testing could produce good results.

Case 2: Direct benefit transfer—a game-changer for financial inclusion in India

History is replete with evidence of the role played by monetary and financial systems in people's lives. Observers describe "golden age" periods in which the quality of life was better, the spirit of innovation encouraged and prosperity widespread. To the extent that this was indeed the case, it was so because of wider participation in formal financial systems.

Today, the extent of such participation is an indicator of economic well-being made possible through opportunities in savings, payments, investments and insurance. Financial services also help small businesses grow, create employment and contribute to economies.

Yet a reported 2.5 billion adults worldwide are excluded from formal financial systems. In India, nearly half of the country's 1.25 billion people have no access to banking services, and close to 300 million people live below the official poverty line. To realize India's potential, its leaders need to ensure that this excluded segment can take part in formal banking and benefit from it.

The majority of people who live below the poverty line—as well as just above it—in India rely on government welfare payments such as those from the National Rural Employment Guarantee Act, old age pensions, scholarships, widows' pensions, discounted liquid petroleum gas (LPG) for cooking, and other subsidies. The government of India makes these payments, whose estimated value is a huge $60 billion per year, through the Direct Benefit Transfer (DBT) scheme, covering beneficiaries under 59 individual schemes.

The sheer extent of its reach and impact makes DBT an excellent platform to bring banking closer to the unbanked masses, but achieving that objective requires long-term vision, not short-term or ad hoc measures.

Over the last decade, technological developments have fast-tracked the way we do financial transactions. Financial inclusion efforts, too, have felt the impact of emerging technologies, with biometrics leading the way in extending banking reach. Technology-aided DBT delivery is the way forward.

However, the issues plaguing the existing transfer system need to be addressed, including broadband and mobile connectivity, lack of power, inaccurate data on beneficiaries, leaks and frauds due to corruption, and so on. This is where technology-driven solutions come into the picture.

JAM (Jan Dhan, Aadhaar, Mobile) is a combination of three platforms intended to drive financial inclusion efforts as envisaged by the government of India. The first is Jan Dhan Yojana, a project for opening bank accounts for all. The second is Aadhaar, a 12-digit unique identity number issued by the Unique Identification Authority of India. And the third is mobile technology, especially the increasing adoption of smartphones. The JAM trinity, as it is called, has the potential to completely digitize DBT delivery and promote financial inclusion.

Under this digital program, the first scheme to be rolled out using DBT is the LPG subsidy payment to nearly 150 million registered beneficiaries. Welfare payments for the National Rural Employment Guarantee Act and kerosene subsidy are expected to roll out next. Intended beneficiaries get their welfare payments directly credited to a bank account and can do financial transactions from multiple access points as is convenient—ATM, bank branch or business correspondent point—with the unique identity number for authentication (see Fig. 10.4).

Source: National Payments Corporation of India. Infographics by FINO PayTech.

The latest government reports say that the implementation of DBT for LPG subsidy beneficiaries has led to savings of more than $3 billion during the last two years. This was possible because of two primary reasons: weeding out over 30 million fake beneficiaries using the unique identity number, and ensuring that the right beneficiary gets the right amount through the safety of a bank account.

If a single scheme can realize such savings, imagine the potential when $60 billion in welfare payments is seamlessly and directly credited to the bank accounts of all beneficiaries. Gradually, as hitherto unbanked and under-banked people start using bank accounts to make transactions, various other relevant financial products can be made available, allowing them to save, invest and insure their future. This innovative model can thus be a game-changer in promoting financial inclusion and fueling economic growth in India.

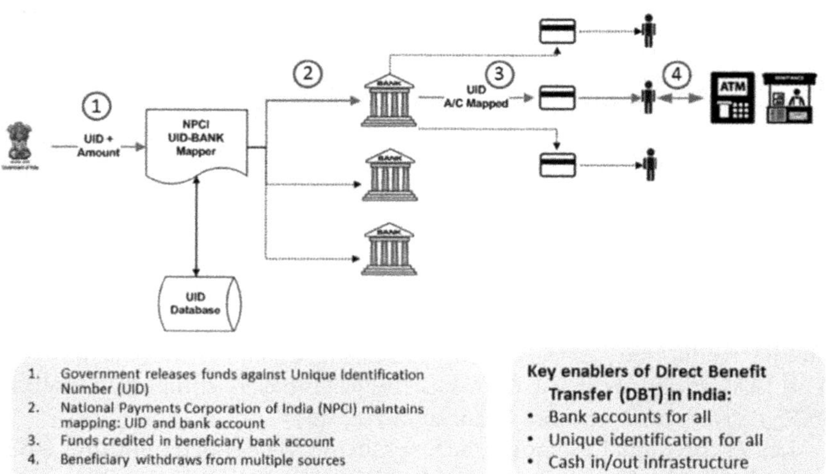

Fig. 10.4 Step direct benefit transfer delivery model. Source: National Payments Corporation of India. Infographics by FINO PayTech

Conclusions

In the age of the digital economy, digital finance with its unique features has been playing an important role in countries' inclusive growth process. It extends the opportunity for all, including the poor and disadvantaged, to participate in economic activities. This study has explored the effects of digital finance on financial inclusion by two methods. Via the first method, using data from 2004 to 2016 on 189 countries, the authors conducted trend analysis and found that all digital finance proxies have positive effects on economic growth. For the second method, considering India as a highly promising country for digital finance, we studied two cases and observed the positive effects of digital finance on financial inclusion. Thus, based on the findings, we argue that policy makers should adopt such policies to promote digital finance, which in turn will promote financial inclusion, resulting in the inclusive growth of the country.

Depending on the availability of data, future research could attempt to examine the relation between digital finance and financial stability with a special focus on periods of financial crisis, so that the economy will have the capacity to meet the challenge of such crises. Digital finance may not serve as an overall panacea for the world's poor, but it can definitely facilitate the better establishment of services to meet working capital requirements, thereby contributing to a corruption-free system leading to better dissemination of services. Predictive analytics combined with biometrics could be one of the futuristic areas of digital finance as the discipline matures and includes more stakeholders across the value chain, making it more inclusive and holistic. The future of digital finance and its implementation is here to stay and would provide ample opportunities for future researchers to expand its aim and scope.

References

Dhawan, S. (2017, September 29). This savings account doesn't require any minimum balance: Find out if it suits you. *Economic Times*, India.

Gabor, D., & Brooks, S. (2017). The digital revolution in financial inclusion: International development in the fintech era. *New Political Economy, 22*(4), 423–436.

Ghosh, S. (2016). Does mobile telephony spur growth? Evidence from Indian states. *Telecommunications Policy, 40*(10–11), 1020–1031.

Gomber, P., Koch, J. A., & Siering, M. (2017). Digital Finance and FinTech: Current research and future research directions. *Journal of Business Economics, 87*(5), 537–580.

GPFI (2010). G20 Principles for Innovative Financial Inclusion—Executive Brief. Retrieved from http://www.gpfi.org/publications/g20-principles-innovative-financial-inclusion-executive-brief

Hannig, A., & Jansen, S. (2010). Financial inclusion and financial stability: Current policy issues. *ADBI Working Paper No. 259*, Asian Development Bank Institute, Tokyo.

Klapper, L., El-Zoghbi, M., & Hess, J. (2016). Achieving the sustainable development goals. The role of financial inclusion. Retrieved May 23, 2016, from http://www.ccgap.org

Manyika, J., Lund, S., Singer, M., White, O., & Berry, C. (2016). *Digital finance for all: Powering inclusive growth in emerging economies*. Washington, DC: McKinsey Global Institute.

Ozili, P. K. (2018). Impact of digital finance on financial inclusion and stability. *Borsa Istanbul Review, 18*(4), 329–340.

Peake, C. (2012). New frontiers: Launching digital financial services in rural areas. In *The 2012 Brookings Blum Roundtable Policy Briefs*. Mercy Corps.

Sarma, M. (2015). Measuring financial inclusion. *Economics Bulletin, 35*(1), 604–611.

Siddik, M. N. A., Sun, G., & Kabiraj, S. (2015). Financial inclusion and its determinants: A study of Bangladesh. *Indian Journal of Finance, 9*(6), 7–29.

CHAPTER 11

Understanding Consumer Behavior in Technology-Mediated Spaces

İlayda İpek

Substantial improvements in technology have been influencing not only strategies and structures of firms but also consumers' social and economic lives. In terms of the former, in order to respond to the alterations in the technological environment, and ultimately, to stay ahead of the competition, firms have been making continuous attempts to fit their internal resources and competencies with the external circumstances. Concerning the latter, having an enormous impact on the manner of communicating, consuming, and progressing, technological advancements have been shaping consumers' daily lives. Specifically, digital technology has affected consumers' routines and habits, which in turn has led to changes in consumer behavior.

Currently, along with the benefits of the Internet, consumers have the ability to easily reach all information regarding firms, which increases the bargaining power of consumers. They have had a more tendency to shop online; however, the critical thing is that they desire to have unique experiences, which stresses the significance of effective customer relationship management (CRM) activities. Being aware of the strength of technology

İ. İpek (✉)
Faculty of Business, Dokuz Eylül University, İzmir, Turkey
e-mail: ilayda.gungor@deu.edu.tr

B. George, J. Paul (eds.), *Digital Transformation in Business and Society*, https://doi.org/10.1007/978-3-030-08277-2_11

on consumer behavior, firms have been more inclined to enhance their offerings, with a particular emphasis on the design of positive consumer experiences. Additionally and remarkably, together with the advances in technology, the roles of firms and consumers have differed considerably. In this sense, firms have been working interdependently with their customers, who are highly involved in co-creating value. Correspondingly, digital marketing has gained huge importance in every aspect of business activities. Building on the aforementioned issues, by drawing on digital marketing, the main purpose of this chapter is to identify how consumer behavior has evolved through the digital transformation in recent decades.

Digital Marketing

Dating back to the early 1990s, the Internet has been commercially utilized since almost three decades (Tiffin & Rajasingham, 2003). Coupled with this digital evolution, boundaries across firms and cultures have been seriously decreased (Wymbs, 2000). In today's business world, rivalry among firms is not based on products or services, but instead on business models, which aim to generate customer value by the combination of digital technologies (Yoffie & Cusumano, 1999). Nevertheless, maintaining competitive advantage in a marketplace is much more burdensome than before by reason of high business risks related to the extremely dynamic nature of the digital environment (Hamel & Sampler, 1998). As it has been well emphasized by Bill Gates (1999), "the Internet changes everything".

Increase in the availability of the Internet; and thus, its extensive commercial usage, have led new players to stand out in today's modern economy. Those players have recorded huge success; such that, according to Forbes (2018a), Google, Facebook, and Amazon have been among the world's five most valuable brands in this year. E-commerce has been dramatically growing; for instance, current retail tendencies indicate that more than half of Americans (51%) have been engaged in online shopping in 2018, with the e-commerce growth rate of 23% year-over-year (BigCommerce, 2018). In addition, it has been expected that US consumers' spending on online shopping will total $632 billion by 2020 (Business Insider Intelligence, 2016).

Moreover, as another distinctive alteration in the technological environment, Web 2.0 has begun to replace Web 1.0. Whereas the contents of Web 1.0 tools are created by providers, Web 2.0 technologies incorporate interactive and user-generated content (Munar & Jacobsen, 2014). In conjunction with the widespread acceptance of Web 2.0 technologies, a substantial switch from firms to consumers has been reflected (Berthon, Pitt, Plangger, & Shapiro, 2012). In spite of its technological nature, Web 2.0 has also yielded to sociological revolutions, which have exerted an enormous influence on business operations (Berthon et al., 2012). Relying on these facts, in recent years firms have been more inclined to enhance "digital relationships" with their customers (Phillips, 2015).

In order to keep pace with the digital transformation of consumer behavior, firms have been oriented to reshape and restructure their functions, with a special stress on marketing operations (Edelman & Heller, 2015). Accordingly, over the past decades, marketing scholars and practitioners have registered a significant change in the nature of marketing, specifically due to the emergence of new marketing communication channels derived from technological innovations (Lamberton & Stephen, 2016). Currently, digital technology has critically altered the way firms formulate their targeting strategies and interact with their customers. In this respect, "digital marketing" arises as a crucial concept.

The digital marketing term is defined as "an adaptive, technology-enabled process by which firms collaborate with customers and partners to jointly create, communicate, deliver, and sustain value for all stakeholders" (Kannan & Li, 2017, p. 23). Digital marketing represents a novel perspective, not only traditional marketing reinforced by digital components; hence, it includes idiosyncratic characteristics that need to be in-depth evaluated (Taiminen & Karjaluoto, 2015). Instead of concentrating on technology, an effective digital marketing strategy necessitates to comprehend how consumers adopt technology, and correspondingly, to look for channel alternatives in an attempt to strengthen the engagement between consumers and brand (Ryan & Jones, 2009). Without a proper analysis, digital marketing communications may not make sense for target customers, which in turn may result in big losses. Dealing with change and innovation seems to be problematic for all marketing practitioners; however, the situation is notably difficult for digital marketers (Chaffey, 2010).

Being aware of the role of digital marketing in consumers' decision making process, in today's business world, it is imperative for firms to apply digital marketing solutions in an effort to reach their target customers. This is especially valid for both large and small- and medium-sized enterprises to stay competitive in market (Taiminen & Karjaluoto, 2015). Newly, digital advertising expenditures provide evidence on this notion;

digital ad spending worldwide accounted for $283.35 billion in 2018, with a growth rate of 21.4% (eMarketer, 2019). In addition, it has been also anticipated that worldwide spending on digital advertising will amount to $517.51 billion by 2023 (eMarketer, 2019). Critically, the amount of digital touchpoints is rising by more than 20% in every year, because more number of consumers prefers to use digital technologies and "younger, digitally oriented consumers enter the ranks of buyers" (Bughin, 2015). Contacting with consumers by means of digital media is therefore recognized as one of the most promising approaches in the area of strategic marketing for the coming decades (Smith, 2011).

Accompanied by digitization, traditional marketing communication tools started to be supported by new digital channels. These digital tools, providing new and exciting platforms, enable firms to reach target markets in various and creative manners (Ryan & Jones, 2009). Importantly, besides its impact on targeting strategies, digital marketing also allows to leverage existing brand equity (Chaffey, 2010). Augmented reality practices, social media, and mobile apps constitute some late communication channels in digital marketing, which have been broadly preferred by today's success-oriented firms. As an illustration, Uber symbolizes an achievement story, mainly by virtue of the right mixture of targeting and digital media. Operations of the firm are managed with a mobile app and the firm has largely employed online advertising in social media to cultivate its business activities. Presently, Uber announced $48 billion valuation (Donnelly, 2018), which signifies an outstanding accomplishment obtained in just nine years. Another similar example in the sharing economy focusing on digital marketing is Airbnb founded in 2008, with the reported value of $31 billion as of March 2017 (Forbes, 2018b). In a nutshell, it has been strongly suggested that "technology has facilitated novel market behaviors, interactions, and experiences" (Lamberton & Stephen, 2016, p. 146).

It is a foregone conclusion that digital marketing strategies are of vital importance in today's highly dynamic environment; nevertheless, it has been also firmly advocated that firms should be careful about the decisions regarding digital marketing channels. Specifically, firms cannot perform in all digital communication channels, because "being active" is the first and foremost factor of success (Kaplan & Haenlein, 2010). The appropriate digital medium needs to be determined on the basis of customers to be targeted and message to be delivered (Kaplan & Haenlein, 2010). Moreover, the number and engagement degrees of customers in a digital platform remarkably influence performance (Trusov, Bodapati, & Bucklin, 2010). As a result, pros and cons of each communication medium require to be deeply scrutinized before resources are assigned to digital marketing activities.

E-Customers

It has been widely acknowledged that increase in consumers' involvement in digital devices has given rise to changes in consumer behavior. In particular, information technology innovations have affected to a greater extent how consumers respond to different market settings (Lamberton & Stephen, 2016). The Internet usage rates all over the world have been growing in each year; the statistical report offered by the World Bank (2018) demonstrates that 45.79% of the world's population used the Internet in 2016, while this number was 43.08% in the previous year. Even more stunning is the Internet adoption statistics in some developed countries; for example, in the UK, Japan, and United Arab Emirates more than 90% of the country's whole population utilized the Internet in 2016 (World Bank, 2018).

Consumers' rising inclination toward using the Internet and digital channels can be well explained by the principles of the Technology Acceptance Model (TAM) (Davis, 1989) and Diffusion of Innovations (Rogers, 1962). In line with the TAM, since digital technology encompasses both perceived usefulness and perceived ease of use (Davis, 1989), consumers have been more willing to accept the transformations in the technological environment. As for the Diffusion of Innovations, on the basis of the fit between the digital technology and the characteristics, which affect the acceleration of innovation (i.e., relative advantage, compatibility, complexity, trialability, and observability) (Rogers, 1962), consumers have become more interested in engaging in the Internet and digital tools (Rodriguez, Peterson, & Krishnan, 2018; Takacs & Freiden, 1998). Hence, progresses in the Internet and Web-based platforms have cultivated online consumer behavior, which has shaped the nature of consumers' daily activities (Tiago & Veríssimo, 2014).

The essence of the interaction between firms and customers is not like before, since "The Internet is not just another marketing channel; it's not just another advertising medium; it's not just a way to speed up transactions; the Internet is the foundation for a new industrial order" (Hamel & Sampler, 1998). Web 2.0 technologies have dramatically resulted in a shift in value creation and power from firms to customers (Berthon et al., 2012). Currently, customers are more empowered than before because of the advantages associated with the employment of digital devices, covering, information access, product variety, competitive pricing, and convenience (Pires, Stanton, & Rita, 2006; Tiago & Veríssimo, 2014). Without much effort, they are able to spread word-of-mouth information about products/services via social networks even with individuals they do not know (Kannan & Li, 2017). Further, as another concern for digital marketers, customers are becoming increasingly sophisticated and their expectations are incessantly expanding (Udo, Bagchi, & Kirs, 2010).

Customers are actively taking part in different digital communication channels such as social media, web forums, blogs, and mobile apps. Whereas previously for product/service purchases, customers were more relying on worth-of-mouth communication with membership reference groups such as family and friends, today they prefer more to gather information about products/services through online reviews, to compare and contrast different offerings' aspects and prices on websites, and to talk over alternatives through social-networking platforms (Edelman, 2010). In conjunction with customers' high authority in digital world, brand messages formed by firms have lost their effect and the possibility of conversion has declined on average (Bughin, 2015). Realizing their bargaining power on firms, customers want firms to build close bonds with them, rather than to command (Edelman, 2010).

Especially, social media acts a critical role for digital customers; in social media customers have an interaction with brands and they exchange information and experiences with others (Lamberton & Stephen, 2016), which have been considered to have a significant influence on buying behavior (Berthon et al., 2012). It has been revealed that in globe 135 minutes were daily spent in social media in 2017, with an increase from 126 minutes in the prior year (Ward, 2018). And interestingly, the statistics referring to the time devoted for mobile devices have indicated that in the US individuals consume more than 90% of their Internet time on smartphone apps (eMarketer, 2017).

In some cases, customers' high involvement in digital mediums may be seen as problematic in terms of firms. As it has been already highlighted, today customers are identified as "unfettered" (Nunes & Cespedes, 2003), and correspondingly, they are free to spread positive and negative words about a firm's offerings and strategies with or without getting permission of the firm (Kietzmann, Hermkens, McCarthy, & Silvestre, 2011). This makes firms defenseless as their offerings' quality, availability, and prices are completely obvious and any weak points are swiftly disseminated all over the world (Nunes & Cespedes, 2003). Such kinds of online customer-to-customer information exchanges are attempted to be managed by firms with word-of-mouth marketing methods such as social media marketing, viral marketing, and guerilla marketing (Kozinets, de Valck, Wojnicki, & Wilner, 2010), because customers are highly dependent on other customers' opinions and feelings while making buying decisions.

Online Customer Experiences

Developments in technology; logistics, payment systems, and trust in particular—together with the escalation in the Internet usage—have formed a $1.9 trillion global online shopping market, in which "millions of consumers no longer 'go' shopping, but literally 'are' shopping - at every moment and everywhere" (KPMG, 2017, p. 2). Primarily due to the benefits provided by online retailing to consumers such as freedom of shopping in any time and place, online sales have smoothly reached double-digit growth numbers in some countries (Nielsen, 2017). Notably, in the world's biggest online market, China, the growth rate of 32.2% year-on-year in online retailing was recorded in 2017, while this number decreased by 2.1% in the first two quarters of 2018 (National Bureau of Statistics of China, 2018a, 2018b). The nature of online shopping is assumed to alter at an even faster speed coupled with the expansion of voice search and omni-platform/omni-channel alternatives (Popomaronis, 2017).

Apart from these online shopping trends, it is indisputable that consumers are not rational decision makers anymore. During the stages of the decision making process, they are more relying on their feelings and they want to have unique experiences. Accordingly, the present situation is more challenging for firms, since they have to devote more time and effort to improve their offerings through well-designed and positive customer experiences. This fact is not only true for brick-and-mortar businesses; by virtue of highly competitive environment, online operations need to be also strengthened by favorable experiences in order to leverage conversion rates. However, this is much more painful, as providing and managing "a balanced experience" on a digital platform is severely complicated (Chaffey, 2010).

Consumers collect information about products/services either by direct experience, such as trial, or by indirect experience, such as exposure to advertising (Li, Daugherty, & Biocca, 2001). Consumer experiences in digital environment are considerably different; such that, this concept has been viewed as indirect experiences; but it also includes social and interpersonal cues (Liao & Keng, 2014). Moreover, in Web-based platforms consumers rest solely on sight and sound, whereas in traditional marketplace experiences encompass all senses (Straker, Wrigley, & Rosemann, 2015). In this regard, online consumer experiences are described as psychological and emotional conditions faced by consumers, meanwhile having interaction with products online (Li et al., 2001). Widespread preference of online shopping has resulted in the evolution of the e-customers' buying behavior, allied with the attainment of e-purchasing experiences (Hernandez, Jimenez, & Martín, 2009).

Increase in the strength of online consumer behavior has caused firms to expand their digital existence (Nielsen, 2017); and hence, to look for the ways to create a unified experience across various channels. On the other hand, it has been shown that one of the most important obstacles of online shopping is the lack of delightful experiences and social interplay (Barlow, Siddiqui, & Mannion, 2004). Consumers remark that online transactions are not specific to a certain customer and they do not feel comfortable while purchasing unusual and intricate products; they therefore expect firms to offer customer assistance not only in offline but also in online environments (Holzwarth, Janiszewski, & Neumann, 2006), that places a high emphasis on the significance of unified customer experience. In this respect, customer experiences have been recognized as the pivotal drivers of online shopping behavior (Childers, Carr, Peck, & Carson, 2001).

Further, as a consequence of the diffusion of digital marketing tools, currently, customers have become more engaged with brands; for example, they receive regular e-mails about new season products, they utilize mobile apps to take the advantage of discount prices, or they visit the brand's social media pages to learn about the upcoming events (Edelman, 2010). Nevertheless, considering customers' involvement in different digital channels at once, firms should be more inclined toward generating multi-channel experiences (Straker et al., 2015). This approach has been thoroughly underlined by the term of "perfect customer experience", which is firmly intertwined with the firm's capability "to gather and deploy customer information from all channels and to integrate it with other relevant information" (Payne & Frow, 2005, p. 173). In this sense, digital channels help firms consolidate customer experiences, which in turn produce high online traffic and superior customer loyalty at the end (Edelman, 2010).

Currently, many firms have leaned toward the integration of online and offline customer experiences. L'Oreal is one of those firms which successfully creates unique experiences for its customers through the combination of online and offline aspects. As an illustration, L'Oreal's Makeup Genius app offers opportunity for customers to virtually try various kinds of makeup products through a webcam (Edelman & Heller, 2015). As another example, last year IKEA introduced IKEA Place app—an augmented reality application, which aims to form memorable customer experiences by enabling people to virtually design their homes, offices, and so on with different furniture (IKEA, 2017).

Furthermore, in recent decades retailing industry has also registered substantial alterations, mainly due to the improvements in online mediums and continuous digitalization (Verhoef, Kannan, & Inman, 2015). In an effort to respond to changes in retailing, firms have begun to simultaneously operate both in traditional and online channels; however, at this point the critical thing is that those retailing activities require to be carried out in an integrated fashion. Assuming that retailing operations are dependent on autonomous systems, fragmented supply chains emerge, which fail to create a coherent and trustworthy customer experience (Saghiri, Wilding, Mena, & Bourlakis, 2017). Concordantly, in the latest years, a shift from multi-channel retailing to omni-channel retailing has been noticed (Verhoef et al., 2015).

Whereas multi-channel retailing fundamentally concentrates on retail channels, in omni-channel retailing the stress is placed on the integration between channels and brands (Picot-Coupey, Huré, & Piveteau, 2016). Symbolizing a challenge for traditional retailers (Rigby, 2011), omni-channel retailing is defined as "the synergetic management of the numerous available channels and customer touchpoints, in such a way that the customer experience across channels and the performance over channels is optimized" (Verhoef et al., 2015, p. 176). It puts a particular emphasis on the interplay of "disparate channels into a single seamless omni-channel experience" (Rigby, 2011, p. 67). The concept has gained a considerable importance in retailing together with increase in customers' demands for unified brand experiences. Concisely, the effective management of the multi-channel customer experiences is particularly vital for outstanding business success and performance in today's highly digital environment.

E-Customer Relationship Management

Relationship marketing, focusing on creating, expanding, and sustaining sufficient relational exchanges, has been seen as a big transformation for both marketing scholars and practitioners (Morgan & Hunt, 1994). It has been strongly suggested that establishing long-term relationships with internal and external partners generates more positive returns compared to discrete transactions (Morgan & Hunt, 1994); that view has caused a fundamental shift from a transaction-based perspective to a relationship-centered orientation. In this regard, within the context of CRM, in what manner customers should be dealt to establish effective bonds with them has become a crucial concern in marketing theory and practice (Reinartz, Krafft, & Hoyer, 2004).

Drawing on the relationship marketing theory, CRM rests primarily on the notion that in order to be successful and survive in the market, instead of looking for ways to sell products/services, firms should consistently direct themselves toward building high customer value, which in turn creates value for the firm (Boulding, Staelin, Ehret, & Johnston, 2005). The main tenets of CRM defend that as long-term customers are more profitable than short-term ones, firms should maintain their relationships with customers to improve customer retention and loyalty (Zeithaml, Berry, & Parasuraman, 1996). Correspondingly, firms have increasingly started to appreciate that the economic worth of customers to the firm is not the same; and thus, they need to tailor their products/services and marketing communications on the basis of various customers (Reinartz et al., 2004).

Even though CRM is frequently adopted to refer to technology-oriented customer solutions (e.g., sales force automation), the breadth of the concept is wider; such that, CRM covers a strategic vision with a particular stress on the interrelationship of organizational culture, information technology, and customer service (Payne & Frow, 2005). Nevertheless, whereas CRM comprises diverse types of applications, a technology-based view is widely accepted (Keramati, Mehrabi, & Mojir, 2010). The fast progress in information technology has allowed firms to adopt new technology-based tools, entitled CRM technology, which aims to support the process of CRM (Jayachandran, Sharma, Kaufman, & Raman, 2005). Currently, many firms make huge investments in CRM technology in an attempt to effectively manage their CRM activities (Powell, Noble, Noble, & Han, 2018).

In the increasingly digitalized world, CRM technology is of great importance; CRM was identified as the biggest software market in 2017, with the worldwide revenue totaling $39.5 billion (Gartner, 2018). In addition, it has been estimated that CRM will grow by 16% in 2018 and will therefore represent the fastest expanding software market in the same year (Gartner, 2018). Adoption rates also signify the strength of CRM in current topics of marketing; with approximately 47% in 2017, on average (Lazar, 2017). Firms prefer to employ CRM technology for different purposes such as more accurate customer profitability evaluation, integrated customer data management, and superior customer lifetime value (Powell et al., 2018). However, albeit its popularity, the consequences of CRM technology are questionable in terms of either customers or firms (Jayachandran et al., 2005).

By virtue of the substantial effect of technological advancements on business processes, CRM operations have been also influenced by Internet-based technologies in a considerable way. In recent decades, an increasing integration between CRM and Internet technology has been noticed, that formed a novel approach for CRM strategy. Being called as electronic CRM (e-CRM), traditional CRM applications have been replaced with a more interactive nature, depending fundamentally on regular and effective customer communication (Kımıloğlu & Zaralı, 2009). Together with exploiting the benefits of Internet technology, e-CRM enables to expedite the handling of customer relationships (Harrigan, Ramsey, & Ibbotson, 2008).

A comprehensive definition of e-CRM involves "the marketing activities, tools and techniques, delivered over the Internet (using technologies such as Web sites and e-mail, data-capture, warehousing and mining) with a specific aim to locate, build and improve long-term customer relationships to enhance their individual potential" (Lee-Kelley, Gilbert, & Mannicom, 2003, p. 241). In this sense, e-CRM offers important advantages in respect of scrutinizing customer data, since at the right time it signals what customers really expect (Harrigan et al., 2008). Thanks to this attribute, it prevents resource losses in production operations systems (Harrigan et al., 2008). Importantly, by means of e-CRM, firms are more able to personalize their bonds with customers, notably owing to higher number of customer touchpoints and greater opportunity to customize marketing communications (Day & Hubbard, 2003).

Referring to the reality of digital consumer behavior and the significance of e-CRM with respect to both efficiency and effectiveness, today firms have been more apt to apply online CRM solutions to attain competitive advantage. Besides, in parallel with the late trends in consumer behavior and the growing impact of Web 2.0 technologies, within the context of e-CRM, new concepts also emerged. Among these, social CRM (or CRM 2.0) serves as a popular philosophy that combines social media and dynamic capabilities of traditional CRM (Greenberg, 2010). This new concept has been recognized as a more cooperative and network-driven perspective to deal with customer relationships (Trainor, Andzulis, Rapp, & Agnihotri, 2014).

Customers have become highly involved in social media channels not only to keep in touch with their friends or colleagues but also to interact with firms (Trainor, 2012). In this regard, the brief purpose of social CRM is to respond to the expectations of social customers (Greenberg, 2010). Via social CRM applications firms endeavor to cultivate their bonds with customers in social networks both with regard to quality and quantity; and thus, to reinforce brand image and overall brand loyalty (Wongsansukcharoen, Trimetsoontorn, & Fongsuwan, 2015). Seen as a new breath in CRM, social CRM is of great benefit to firms. such that it has been found that customer retention rates increase by 26%, thanks to social CRM applications (Lazar, 2017). Considering these facts, traditional CRM service providers (e.g., Salesforce.com and SAP) have begun to merge data from social media channels such as Facebook and Twitter to actively monitor customers and quickly respond to customer requirements (Owyang, 2010).

Customer Co-Creation

Traditional value creation approach is based on the thought that firms establish value and transfer it to customers (Prahalad & Ramaswamy, 2004a). Specifically, with limited or without interaction with customers, firms independently decide on which products to be manufactured, which marketing messages to be delivered, which distribution channels to be used, and so on so forth (Prahalad & Ramaswamy, 2004b). On the other hand, fundamentally because of the upheavals in customers' roles in recent decades, the main tenets of the traditional value creation view started to be doubted (Prahalad & Ramaswamy, 2004a). It has been suggested that this passive perspective regarding customers should be changed with an active approach, since the real value of a product/service can be best reflected in the eyes of customers (Witell, Kristensson, Gustafsson, & Löfgren, 2011).

Coupled with the big transformation in the essence of the relationship between customers and firms, the process of value building has considerably altered; currently, customers have tended toward taking an active part in each step of firms' value creation processes (Prahalad & Ramaswamy, 2004b). Being described as "co-creation" of value, customers have begun to have an impact not only on product development stages but also on service designs (Heidenreich, Wittkowski, Handrich, & Falk, 2015). Based on this, it has been contended that marketing has recorded a significant change from a goods-dominant logic, focusing mainly on "tangible output and discrete transactions", to a service-dominant logic, which depends strictly on "intangibility, exchange processes, and relationships" (Vargo & Lusch, 2004, p. 2).

Rather than following the traditional definitions of the term, the service-dominant approach identifies services as "the application of specialized competences (knowledge and skills) through deeds, processes, and performances for the benefit of another entity or the entity itself" (Vargo & Lusch, 2004, p. 2) and it places special emphasis on the collaboration between the firm and its stakeholders, which are highly involved in the co-creation of value (Lusch, Vargo, & O'Brien, 2007). This perspective claims that customers represent stakeholders, who consistently make contributions to value generation, whereas firms refer to enterprises with an opportunity to only propose value (Vargo & Lusch, 2004).

Growing engagement in the usage of the Internet and social media channels has facilitated the implementation of co-creation activities. Accordingly, in conjunction with the benefits of digital technology, firms have been increasingly formulating successful co-creation strategies. For instance, Dell's Ideastorm exemplifies one of those impressive applications in value co-creation (Saarijärvi, Kannan, & Kuusela, 2013). In this platform, customers of Dell express their opinions and suggestions about how to improve the offerings of the firm. Owing to this, Dell has an access to customers' actual ideas in relation to Dell's products, and at the same time customers see that their contributions are of great prominence for Dell, that is likely to strengthen customer participation and citizenship behavior (Yi & Gong, 2013).

In this context, Threadless.com also sets a good example (Saarijärvi et al., 2013). The firm calls its customers to design different range of products, covering t-shirts, accessories, and home decor, and so on, and to load their drafts on its website. All customers can give scores for the sketches and total customer scores determine which products will be sold under the name of Threadless.com. By means of this co-creation of value, customers are able to purchase their favorite products and the firm can anticipate future customer demands (Saarijärvi et al., 2013). Furthermore, BMW also cooperates with its customers to put its cars into final form (BMW Group, 2010). In particular, the firm introduced the Co-Creation Lab, which serves as a virtual gathering place where customers are encouraged to comment on automotive world in the future. BMW utilizes the sharing of the customers in order to plan the upcoming projects.

Building on the foregoing, in an effort to exploit the advantages of value co-creation, firms have been trying their best to employ customer resources in an efficacious manner. By means of co-creation activities, firms collect information about future customer requirements and expectations (Gustafsson, Kristensson, & Witell, 2012), which cultivate open innovation (Martini, Massa, & Testa, 2014). By virtue of this close interaction with customers in value creation, firms are able to achieve a competitive positional advantage both in terms of efficiency and effectiveness (Grissemann & Stokburger-Sauer, 2012). The benefits of value co-creation are not confined to firms; but also include customers (Dong, Evans, & Zou, 2008). For example, co-creation of value allows customers to enjoy getting quick response to their demands and having products/services with lower prices (Dong et al., 2008).

Conclusion

In the light of the aforementioned issues, considerable developments in technology have been exerting an enormous impact on both firms and consumers. From the firms' standpoint, necessary adjustments in organizational structures and strategies need to be made in order to keep pace with the latest technological progress; however, more critically, firms should be simultaneously involved in scrutinizing deeply and understanding ever-changing consumer behavior. This is notably because in today's highly digitalized world the way consumers use products/services, communicate with their environment, and interact with firms has been experiencing a huge change. In this sense, tremendous alterations in consumer behavior have been putting a great pressure on firms to be more leaned toward utilizing novel marketing programs and solutions in general, and digital marketing activities in particular.

Importantly, together with the increase in the extension rate of information and communication technologies, the authority of market knowledge has started to be passed from suppliers to consumers, which has yielded to the transformations of consumer empowerment, and ultimately, marketing implications (Pires et al., 2006; Tiago & Veríssimo, 2014). Currently, e-consumers are more able to easily compare and contrast features, prices, and varieties of products/services offered by various firms and even more significant to contact with firms and other consumers without difficulty. Hence, consumers have been considered being more sophisticated and holding more bargaining power than before (Berthon et al., 2012; Udo et al., 2010), which puts special emphasis on the fact that digital engagement between firms and consumers has become much more prominent.

Specifically, relying on the notion that in conjunction with rational drivers, consumers take emotional factors into consideration during their decision making processes (Schmitt, 1999), a vital requirement emerges to provide positive and unique customer experiences not only through offline but also through online platforms in the contemporary marketplace. However, at this point, a critical concern that warrants attention by marketing practitioners pertains to the design of unified customer experiences which balance and harmonize experiences across different marketing channels. In this respect, omni-channel retailing arises as a valuable concept, of which applications are promising in terms of the management of customer experiences, since "the distinctions between physical and online will vanish, turning the world into a showroom without walls" (Brynjolfsson, Hu, & Rahman, 2013, p. 2).

Additionally, digital consumer behavior has also reflected in the revolution of value creation perspective; such that, customers have increasingly stood out as co-creators of value. In recent decades, the collaboration between firms and customers has been much greater and customers have been taking an active part more in establishing value, echoing successful real life examples. As customers represent the best sources of information regarding a value of a product/service (Witell et al., 2011), firms should be more inclined to work together with customers in an attempt to enhance their offerings and to gain a competitive advantage in the marketplace. Drawing on all of these important evolutions in consumer behavior, e-CRM applications have become even more paramount, which underline the significance of creating long-term and effective bonds with digital customers. All in all, this chapter spotlights how consumer behavior has remarkably changed on the basis of the digital metamorphosis, with the particular stress placed on the main tenets of digital marketing and e-customers.

References

Barlow, A. K., Siddiqui, N. Q., & Mannion, M. (2004). Developments in information and communication technologies for retail marketing channels. *International Journal of Retail and Distribution Management, 32*(3), 157–163.

Berthon, P. R., Pitt, L. F., Plangger, K., & Shapiro, D. (2012). Marketing meets Web 2.0, social media, and creative consumers: Implications for international marketing strategy. *Business Horizons, 55*(3), 261–271.

BigCommerce. (2018). The 19 ecommerce trends + 147 online shopping stats fueling sales growth in 2018. Retrieved September 6, 2018, from https://www.bigcommerce.com/blog/ecommerce-trends/

BMW Group. (2010). BMW Group Co-Creation Lab. Retrieved September 27, 2018, from https://www.press.bmwgroup.com/global/article/detail/T0082655EN/bmw-group-co-creation-lab?language=en

Boulding, W., Staelin, R., Ehret, M., & Johnston, W. J. (2005). A customer relationship management roadmap: What is known, potential pitfalls, and where to go. *Journal of Marketing, 69*(4), 155–166.

Brynjolfsson, E., Hu, Y.J., & Rahman, M.S. (2013). Competing in the age of omnichannel retailing. *MIT Sloan Management Review*, Summer, 1–7.

Bughin, J. (2015). Brand success in an era of Digital Darwinism. *McKinsey Quarterly.* Retrieved September 9, 2018, from https://www.mckinsey.com/industries/high-tech/our-insights/brand-success-in-an-era-of-digital-darwinism

Business Insider Intelligence. (2016). Retail Forecast: U.S. consumers will spend $632 billion online by 2020. Retrieved September 6, 2018, from https://www.businessinsider.com/ecommerce-online-retail-monetization-strategy-planning-2016-11

Chaffey, D. (2010). Applying organisational capability models to assess the maturity of digital-marketing governance. *Journal of Marketing Management, 26*(3–4), 187–196.

Childers, T. L., Carr, C. L., Peck, J., & Carson, S. (2001). Hedonic and utilitarian motivations for online retail shopping behavior. *Journal of Retailing, 77*(4), 511–535.

Davis, F. (1989). Perceived usefulness, perceived ease of use, and user acceptance of information technology. *MIS Quarterly, 13*(3), 319–340.

Day, G. S., & Hubbard, K. J. (2003). Customer relationships go digital. *Business Strategy Review, 14*(1), 17–26.

Dong, B., Evans, K. R., & Zou, S. (2008). The effects of customer participation in co-created service recovery. *Journal of the Academy of Marketing Science, 36*(1), 123–137.

Donnelly, G. (2018). Travis Kalanick net worth tops $4 billion with sale of Uber stock. *Fortune.* Retrieved September 8, 2018, from http://fortune.com/2018/01/19/travis-kalanick-net-worth/

Edelman, D. (2010). Four ways to get more value from digital marketing. *McKinsey Quarterly.* Retrieved September 9, 2018, from https://www.mckinsey.com/business-functions/marketing-and-sales/our-insights/four-ways-to-get-more-value-from-digital-marketing

Edelman, D., & Heller, J. (2015). How digital marketing operations can transform business. *McKinsey Digital.* Retrieved September 6, 2018, from https://www.mckinsey.com/business-functions/marketing-and-sales/our-insights/how-digital-marketing-operations-can-transform-business eMarketer. (2017). Mobile time spent in 2018: Will smartphones remain ascendant? Retrieved September 7, 2018, from https://www.emarketer.com/content/mobile-time-spent-2018

eMarketer. (2019). Digital Ad spending 2019. Retrieved May 8, 2019, from https://www.emarketer.com/content/global-digital-ad-spending-2019

Forbes. (2018a). The world's most valuable brands. Retrieved September 6, 2018, from https://www.forbes.com/powerful-brands/list/#tab:rank

Forbes. (2018b). As a rare profitable unicorn, Airbnb appears to be worth at least $38 billion. Retrieved September 8, 2018, from https://www.forbes.com/sites/greatspeculations/2018/05/11/as-a-rare-profitable-unicorn-airbnb-appears-to-be-worth-at-least-38-billion/#2c045ff22741

Gartner. (2018). Gartner says CRM became the largest software market in 2017 and will be the fastest growing software market in 2018. Retrieved September 17, 2018, from https://www.gartner.com/newsroom/id/3871105

Gates, B. (1999). Business @ the speed of thought. *Business Strategy Review, 10*(2), 11–18.

Greenberg, P. (2010). The impact of CRM 2.0 on customer insight. *Journal of Business and Industrial Marketing, 25*(6), 410–419.

Grissemann, U. S., & Stokburger-Sauer, N. E. (2012). Customer co-creation of travel services: The role of company support and customer satisfaction with the co-creation performance. *Tourism Management, 33*(6), 1483–1492.

Gustafsson, A., Kristensson, P., & Witell, L. (2012). Customer co-creation in service innovation: A matter of communication? *Journal of Service Management, 23*(3), 311–327.

Hamel, G., & Sampler, J. (1998). The e-corporation more than just web-based, it's building a new industrial order. *Fortune*. Retrieved September 9, 2018, from https://money.cnn.com/magazines/fortune/fortune_archive/1998/12/07/252120/index.htm

Harrigan, P., Ramsey, E., & Ibbotson, P. (2008). E-CRM in SMEs: An exploratory study in Northern Ireland. *Marketing Intelligence and Planning, 26*(4), 385–404.

Heidenreich, S., Wittkowski, K., Handrich, M., & Falk, T. (2015). The dark side of customer co-creation: Exploring the consequences of failed co-created services. *Journal of the Academy of Marketing Science, 43*(3), 279–296.

Hernandez, B., Jimenez, J., & Martín, M. J. (2009). Adoption vs acceptance of e-commerce: Two different decisions. *European Journal of Marketing, 43*(9/10), 1232–1245.

Holzwarth, M., Janiszewski, C., & Neumann, M. M. (2006). The influence of avatars on online consumer shopping behavior. *Journal of Marketing, 70*(4), 19–36.

IKEA. (2017). IKEA launches IKEA place, a new app that allows people to virtually place furniture in their home. Retrieved September 24, 2018, from https://www.ikea.com/us/en/about_ikea/newsitem/091217_IKEA_Launches_IKEA_Place

Jayachandran, S., Sharma, S., Kaufman, P., & Raman, P. (2005). The role of relational information processes and technology use in customer relationship management. *Journal of Marketing, 69*(4), 177–192.

Kannan, P. K., & Li, H. A. (2017). Digital marketing: A framework, review and research agenda. *International Journal of Research in Marketing, 34*(1), 22–45.

Kaplan, A. M., & Haenlein, M. (2010). Users of the world, unite! The challenges and opportunities of social media. *Business Horizons, 53*(1), 59–68.

Keramati, A., Mehrabi, H., & Mojir, N. (2010). A process-oriented perspective on customer relationship management and organizational performance: An empirical investigation. *Industrial Marketing Management, 39*(7), 1170–1185.

Kietzmann, J. H., Hermkens, K., McCarthy, I. P., & Silvestre, B. S. (2011). Social media? Get serious! Understanding the functional building blocks of social media. *Business Horizons, 54*(3), 241–251.

Kımıloğlu, H., & Zaralı, H. (2009). What signifies success in e-CRM? *Marketing Intelligence and Planning, 27*(2), 246–267.

Kozinets, R. V., de Valck, K., Wojnicki, A. C., & Wilner, S. J. (2010). Networked narratives: Understanding word-of-mouth marketing in online communities. *Journal of Marketing, 74*(2), 71–89.

KPMG. (2017). The truth about online consumers: 2017 global online consumer report. Retrieved September 11, 2018, from https://assets.kpmg.com/content/dam/kpmg/xx/pdf/2017/01/the-truth-about-online-consumers.pdf

Lamberton, C., & Stephen, A. T. (2016). A thematic exploration of digital, social media, and mobile marketing: Research evolution from 2000 to 2015 and an agenda for future inquiry. *Journal of Marketing, 80*(6), 146–172.

Lazar, M. (2017). Ecommerce CRM software: 2017 CRM statistics show why it's a powerful marketing weapon. *IBM Community*. Retrieved September 17, 2018, from https://www.ibm.com/developerworks/community/blogs/d27b1c65-986e-4a4f-a491-5e8eb23980be/entry/2017_CRM_Statistics_Show_Why_it_s_a_Powerful_Marketing_Weapon?lang=en

Lee-Kelley, L., Gilbert, D., & Mannicom, R. (2003). How e-CRM can enhance customer loyalty. *Marketing Intelligence and Planning, 21*(4), 239–248.

Li, H., Daugherty, T., & Biocca, F. (2001). Characteristics of virtual experience in electronic commerce: A protocol analysis. *Journal of Interactive Marketing, 15*(3), 13–30.

Liao, T. H., & Keng, C. J. (2014). Online purchase delay: The roles of online consumer experiences. *Journal of Electronic Commerce Research, 15*(2), 133–150.

Lusch, R. F., Vargo, S. L., & O'Brien, M. (2007). Competing through service: Insights from service-dominant logic. *Journal of Retailing, 83*(1), 5–18.

Martini, A., Massa, S., & Testa, S. (2014). Customer co-creation projects and social media: The case of Barilla of Italy. *Business Horizons, 57*(3), 425–434.

Morgan, R. M., & Hunt, S. D. (1994). The commitment-trust theory of relationship marketing. *Journal of Marketing, 58*(3), 20–38.

Munar, A. M., & Jacobsen, J. K. S. (2014). Motivations for sharing tourism experiences through social media. *Tourism Management, 43*, 46–54.

National Bureau of Statistics of China. (2018a). Total retail sales of consumer goods in December 2017. Retrieved September 11, 2018, from http://www.stats.gov.cn/english/pressrelease/201801/t20180126_1577681.html

National Bureau of Statistics of China. (2018b). Total retail sales of consumer goods in June 2018. Retrieved September 11, 2018, from http://www.stats.gov.cn/english/pressrelease/201807/t20180719_1610962.html

Nielsen. (2017). What's in-store for online grocery shopping: Omnichannel strategies to reach crossover shoppers. Retrieved September 11, 2018, from https://www.nielsen.com/content/dam/nielsenglobal/de/docs/Nielsen%20Global%20Connected%20Commerce%20Report%20January%202017.pdf

Nunes, P., & Cespedes, F. V. (2003, November). The customer has escaped. *Harvard Business Review, 81*, 96–105.

Owyang, J. (2010). Social CRM: Not your father's CRM. *Forbes*. Retrieved September 24, 2018, from https://www.forbes.com/2010/04/09/facebook-twitter-social-media-crm-dell-comcast-cmo-network-jeremiah-owyang.html#225e9e303d89

Payne, A., & Frow, P. (2005). A strategic framework for customer relationship management. *Journal of Marketing, 69*(4), 167–176.

Phillips, E. (2015). Retailers scale up online sales distribution networks. *Wall Street Journal*. Retrieved September 6, 2018, from https://www.wsj.com/articles/retailers-scale-up-online-sales-distribution-networks-1447792869

Picot-Coupey, K., Huré, E., & Piveteau, L. (2016). Channel design to enrich customers' shopping experiences: Synchronizing clicks with bricks in an omnichannel perspective—The direct optic case. *International Journal of Retail and Distribution Management, 44*(3), 336–368.

Pires, G. D., Stanton, J., & Rita, P. (2006). The internet, consumer empowerment and marketing strategies. *European Journal of Marketing, 40*(9/10), 936–949.

Popomaronis, T. (2017). E-Commerce in 2018. Here's what the experts are predicting. *Forbes*. Retrieved September 12, 2018, from https://www.forbes.com/sites/tompopomaronis/2017/12/15/e-commerce-in-2018-heres-what-the-experts-are-predicting/#65a9c2596deb

Powell, A., Noble, C. H., Noble, S. M., & Han, S. (2018). Man vs machine: Relational and performance outcomes of technology utilization in small business CRM support capabilities. *European Journal of Marketing, 52*(3/4), 725–757.

Prahalad, C. K., & Ramaswamy, V. (2004a). *The future of competition: Co-creating unique value with customers*. Boston: Harvard Business School Press.

Prahalad, C. K., & Ramaswamy, V. (2004b). Co-creation experiences: The next practice in value creation. *Journal of Interactive Marketing, 18*(3), 5–14.

Reinartz, W., Krafft, M., & Hoyer, W. D. (2004). The customer relationship management process: Its measurement and impact on performance. *Journal of Marketing Research, 41*(3), 293–305.

Rigby, D. (2011). The future of shopping. *Harvard Business Review, 89*(12), 65–76.

Rodriguez, M., Peterson, R. M., & Krishnan, V. (2018). Impact of CRM technology on sales Process behaviors: Empirical results from US, Europe, and Asia. *Journal of Business-to-Business Marketing, 25*(1), 1–10.

Rogers, E. M. (1962). *The diffusion of innovations*. New York: Free Press.

Ryan, D., & Jones, C. (2009). *Understanding digital marketing: Marketing strategies for engaging the digital generation*. London: Kogan Page.

Saarijärvi, H., Kannan, P. K., & Kuusela, H. (2013). Value co-creation: Theoretical approaches and practical implications. *European Business Review, 25*(1), 6–19.

Saghiri, S., Wilding, R., Mena, C., & Bourlakis, M. (2017). Toward a three-dimensional framework for omni-channel. *Journal of Business Research, 77*, 53–67.

Schmitt, B. (1999). Experiential marketing. *Journal of Marketing Management, 15*(1–3), 53–67.

Smith, K. T. (2011). Digital marketing strategies that Millennials find appealing, motivating, or just annoying. *Journal of Strategic Marketing, 19*(6), 489–499.

Straker, K., Wrigley, C., & Rosemann, M. (2015). The role of design in the future of digital channels: Conceptual insights and future research directions. *Journal of Retailing and Consumer Services, 26*, 133–140.

Taiminen, H., & Karjaluoto, H. (2015). The usage of digital marketing channels in SMEs. *Journal of Small Business and Enterprise Development, 22*(4), 633–651.

Takacs, S. J., & Freiden, J. B. (1998). Changes on the electronic frontier: Growth and opportunity of the World-Wide Web. *Journal of Marketing Theory and Practice, 6*(3), 24–37.

Tiago, M. T. P. M. B., & Veríssimo, J. M. C. (2014). Digital marketing and social media: Why bother? *Business Horizons, 57*(6), 703–708.

Tiffin, J., & Rajasingham, L. (2003). *The Global Virtual University*. London: RoutledgeFalmer.

Trainor, K. J. (2012). Relating social media technologies to performance: A capabilities-based perspective. *Journal of Personal Selling and Sales Management, 32*(3), 317–331.

Trainor, K. J., Andzulis, J. M., Rapp, A., & Agnihotri, R. (2014). Social media technology usage and customer relationship performance: A capabilities-based examination of social CRM. *Journal of Business Research, 67*(6), 1201–1208.

Trusov, M., Bodapati, A. V., & Bucklin, R. E. (2010). Determining influential users in internet social networks. *Journal of Marketing Research, 47*(4), 643–658.

Udo, G. J., Bagchi, K. K., & Kirs, P. J. (2010). An assessment of customers' e-service quality perception, satisfaction and intention. *International Journal of Information Management, 30*(6), 481–492.

Vargo, S. L., & Lusch, R. F. (2004). Evolving to a new dominant logic for marketing. *Journal of Marketing, 68*(1), 1–17.

Verhoef, P. C., Kannan, P. K., & Inman, J. J. (2015). From multi-channel retailing to omni-channel retailing: Introduction to the special issue on multi-channel retailing. *Journal of Retailing, 91*(2), 174–181.

Ward, T. (2018). How much social media is too much? *Fortune*. Retrieved September 8, 2018, from https://www.forbes.com/sites/tomward/2018/06/08/how-much-social-media-is-too-much/#299a7b3360e6

Witell, L., Kristensson, P., Gustafsson, A., & Löfgren, M. (2011). Idea generation: Customer co-creation versus traditional market research techniques. *Journal of Service Management, 22*(2), 140–159.

Wongsansukcharoen, J., Trimetsoontorn, J., & Fongsuwan, W. (2015). Social CRM, RMO and business strategies affecting banking performance effectiveness in B2B context. *Journal of Business and Industrial Marketing, 30*(6), 742–760.

World Bank. (2018). Individuals using the Internet (& of population). Retrieved September 7, 2018, from https://data.worldbank.org/indicator/it.net.user.zs

Wymbs, C. (2000). How e-commerce is transforming and internationalizing service industries. *Journal of Services Marketing, 14*(6), 463–477.

Yi, Y., & Gong, T. (2013). Customer value co-creation behavior: Scale development and validation. *Journal of Business Research, 66*(9), 1279–1284.

Yoffie, D. B., & Cusumano, M. A. (1999, January–February). Judo strategy: The competitive dynamics of internet time. *Harvard Business Review, 77*, 70–81.

Zeithaml, V. A., Berry, L. L., & Parasuraman, A. (1996). The behavioral consequences of service quality. *Journal of Marketing, 60*(2), 31–46.

CHAPTER 12

An Evaluation of the National Open Government Data (OGD) Portal of the United Arab Emirates

Stuti Saxena

Introduction

Conceived both as a philosophy and as a set of policies, open government data (OGD) relates to the data published by the government which is freely available for being re-used by different stakeholder groups (citizens, journalists, researchers, entrepreneurs, etc.) to derive value out of the same and improvise upon their services (Alexopoulos, Loukis, & Charalabidis, 2014; Attard, Orlandi, Scerri, & Auer, 2015; Kassen, 2013). Such data may relate to different socio-economic sectors like weather, agriculture, industry, energy, power, education, trade and so on. (Ubaldi, 2013). OGD is an advanced format of e-government which promotes transparency, accountability and citizen trust because the government publishes the hitherto-reserved data for wider usage (Mpinganjira, 2015). Apart from facilitating citizen engagement, OGD initiatives help in improvising upon public policies and contribute toward efficiency,

S. Saxena (✉)
Central University of Haryana, Mohindergarh, India
e-mail: stutisaxena6691@cuh.ac.in

B. George, J. Paul (eds.), *Digital Transformation in Business and Society*, https://doi.org/10.1007/978-3-030-08277-2_12

economy and effectiveness of the government. As such, open data may be defined as "machine-readable data (which is) discoverable, available, and downloadable through dedicated internet portals without cost to potential data users" (Dawes, Vidiasova, & Parkhimovich, 2016, p. 15). For earning benefits from the OGD initiative, it is important that the datasets are published in a timely manner and they should be complete in all aspects (e.g., metadata, data entries, legible formats, provision of visualization and analytical tools, etc.) (Jaeger, Bertot, & Shilton, 2012).

Open data initiatives spur innovation through co-creation activities of the government and the other stakeholders (Verhulst & Young, 2016). For instance, software developers and the government collaborate on the social coding platform GitHub to contribute toward the revamped version of the source code (Mergel, 2015). It has also been underlined that sustainable OGD initiatives help in the economic growth of the country (Charalabidis, Alexopoulos, & Loukis, 2016; Jung & Park, 2015; Saxena, 2017a; Wirtz & Birkmeyer, 2015). Implicitly, the OGD initiative of the United Arab Emirates (UAE) has the potential of facilitating the economic diversification of the country and meeting the goals of the Vision 2021 espoused by the government. However, to assess the potential of the OGD initiative, it is important that the datasets published by the government online are qualitatively and quantitatively adequate. This would facilitate optimum re-use of the datasets for deriving social and economic value out of them.

In line with the aforesaid, this chapter invokes the usability framework (Machova, Hub, & Lnenicka, 2018) and the website quality framework (Sorum, Andersen, & Clemmensen, 2013) to present a case study where we evaluate the national OGD portal of the UAE (https://bayanat.ae/) in terms of the data quality and underline the drivers and barriers in re-using the datasets published via the portal. The key research question guiding the case study is: "What are the drivers and barriers in re-using the datasets published via the national OGD portal of UAE?" Further, the findings from the case study emphasize the need for maintaining the quality of the datasets published therein. As the first detailed study to investigate the national OGD portal of the country, the case study is a significant contribution to the OGD literature.

The chapter is structured as follows: After summarizing the literature on OGD, a brief shall be provided about the research methodology invoked in the chapter wherein the case study on the UAE's national OGD portal shall be elaborated. Thereafter, concluding remarks shall be provided to summarize the major inferences. The penultimate sections of the chapter shall detail the social and practical implications of the case study and leave pointers for further research.

Related Research

OGD is an emerging phenomenon and academic interest in the field has been increasing over a period of time. While most of the studies are conducted in the West, where open data initiatives are being undertaken at an advanced level, studies in the developing countries are few and far between (Huijboom & Van den Broek, 2011; Janssen, Charalabidis, & Zuiderwijk, 2012; Saxena, 2017a; Saxena & Janssen, 2017; Zuiderwijk & Janssen, 2014). Implicitly, there is a need for conducting more research in the developing countries to underline the impediments in the implementation of the OGD initiative (Charalabidis et al., 2016) besides drawing lessons from the countries that have been successful in the implementation of the open data initiatives (Nugroho, Zuiderwijk, Janssen, & de Jong, 2015). Such contextual studies, including the empirical ones, provide cases of the success or failure of OGD initiatives. For instance, Rothenberg (2012) followed a case study approach wherein the OGD frameworks of the US, the UK, Canada and New Zealand were compared. The purpose of selecting these four countries lay in the marked similarities of culture among them. Likewise, a study based in Africa aimed at understanding the extent of OGD implementation regarding the characteristics of an "ideal" OGD portal and the study was conducted in five (Ghana, Kenya, Sierra Leone, South Africa and Tanzania) out of the seven (Ghana, Sierra Leone, Tunisia, Morocco, South Africa, Kenya and Tanzania) OGD centers (Afful-Dadzie & Afful-Dadzie, 2017). In another study, the Unified Theory of Acceptance and Use of Technology (UTAUT) model was adapted in order to conduct an empirical investigation in the Netherlands and identify the factors which impact the usage and adoption of OGD (Zuiderwijk, Janssen, & Dwivedi, 2015). In an empirical investigation conducted on a sample of 210 citizens in Germany, the extended Technology Acceptance Model (TAM) was deployed to underline the relationship between the constructs (ease of use, usefulness, transparency expectancy, participation expectancy, collaboration, intention to use OGD and word-of-mouth intention concerning OGD) and it was found that all the five constructs were significant predictors of intention to use OGD and to collaborate among each other using word-of-mouth (Wirtz, Weyerer, & Rosch, 2017). In another empirical investigation conducted in India with a sample of 244 respondents, OGD use and acceptance among different users was probed and it was concluded that OGD use has increased in the country and men are more likely to tap OGD than women (Saxena & Janssen, 2017). All these studies underlined the reasons as to why OGD initiatives are yet to realize their cherished aims in terms of promoting accountability in administration and furthering citizen participation in policy-making.

Literature on OGD has provided an understanding that the demand-supply equation of any OGD initiative merits a careful consideration and this entails that the legal provisions regarding publication of datasets and the concern for privacy of the individuals (Martin, 2014). For instance, datasets are "supplied" by the government agencies and datasets may also be "suggested" or "recommended" by the users. Likewise, datasets may be "demanded" by the stakeholders for their purposes. The "supply" side may be bolstered by ensuring that the entire OGD publication value-chain is qualitatively and quantitatively superior and the "demand" side may be strengthened by the user engagement and involvement in the OGD initiative.

Many studies have provided frameworks for investigating OGD initiatives. For instance, Martin (2014) has provided a framework for underlining the social and technological aspects of OGD initiative. The framework includes five dimensions—digital technologies (configurations that include tangible artifacts, the skills of technologists and users and the interfaces of artifacts with the wider technical infrastructure), user practices (manner in which data is being re-used by a large cross-section of stakeholders), public management practices (includes the processes of data and ICT management and established data-related policies), institutions (include the sets of rules that connect data users and government organizations, including data markets and regulatory frameworks for government data) and resources (the resources drawn upon by actors shaping the OGD data agenda as including: social capital (the networks that connect actors); cultural capital (cultural goods and services, such as skills and knowledge); economic capital (money and other assets that can be directly and immediately converted to money); and symbolic capital: the means available on the basis of (perceived) prestige or legitimacy.

Another study provides a model which outlines the stages in the roll-out of any OGD initiative (Kalampokis, Tambouris, & Tarabanis, 2011a; 2011b). Kalampokis et al. (2011a) have classified OGD into "downloadable files" (data is available in simple formats), "linked data" (data is linked with another one and re-used), "direct data provision" (all data is available via a portal and synchronized with time) and "indirect data provision" (actual data is provided and the user is responsible for further aggregation and processing of the data). Therefore, a four-stage model (aggregation of government data; integration of government data; integration of government data with non-government data; and integration of government data with non-government formal and social data) is identified by Kalampokis et al. (2011a). However, the drawback of the study lies in being more technical in approach wherein the emphasis was laid down on the crude mechanistic dimensions of linked data.

Saxena (2017b) provided a typology of countries on the basis of their OGD-adherence ("laggard," "caged," "forerunner" and "champ") wherein the "laggard" countries are the ones where there are hindrances associated with OGD implementation and OGD-usage; "caged" countries are those with less propensity to implement an OGD initiative but increased potential of usage by different stakeholders; "forerunner" countries as those which hold high potential of rolling out an OGD initiative but low potential of usage by different stakeholders; and "champ" countries as those which ranked high in terms of implementation of an OGD program as well as usage by a diverse set of stakeholders (p. 219).

In another study, another model expounds the degree of interaction among the stakeholders involved in the OGD initiatives directly or indirectly (Sieber & Johnson, 2015). Thus, Sieber and Johnson (2015) provide four models which deal with the "nature of (OGD) delivery (which) shapes the way the data is used" (Sieber & Johnson, 2015, p. 310). Table 12.1 summarizes the four models.

While the aforementioned models are descriptive, the overall comprehensiveness in terms of quality assessment of the datasets is missing. Therefore, the present case study seeks to adopt the evaluation framework proposed by Machova and his colleagues (2018). In their benchmarking framework derived from the reading of the literature on OGD, three dimensions (criteria) and 14 sub-criteria were defined which were a function of data discoverability, data accessibility and reusability. Table 12.2 reproduces the framework proposed by them.

Apart from this, the case study underlines the significance of maintaining the quality of the website dedicated for the publication of datasets. For this purpose, the website quality framework (Sorum et al., 2013) is being referred here. The basic premise of the framework is that the

Table 12.1 Citizen engagement models proposed by Sieber and Johnson (2015)

Model	*Description*
Data over the wall	This is the basic citizen engagement model wherein the government publishes datasets on the online portal directly. Basic features like downloading of datasets, visualization, mapping or sorting are allowed. Datasets are available in formats like PDF, Excel and so on. Programmatic access may be permissible via software-to-software interface (i.e., application programming interface or API). Users are encouraged to report errors in the form of feedback.
Code exchange	Government encourages the re-use of datasets for innovating products and services. Therefore, the government publicizes and promotes the OGD initiative by holding events, conferences, workshops or "app" contests. Software or application "app" developers, civic hackers and social entrepreneurs are encouraged to participate in such promotional activities.
Civic issue tracker	Citizens contribute to the existing datasets in many ways. For instance, citizens report of civic problems (e.g., fire, accidents, drainage problems, floods, potholes, etc.) which necessitate immediate action by the government authorities. By encouraging participation of citizens in the OGD initiative, the government is regularly informed about the civic issues and this facilitates in revising and updating the datasets (Alexopoulos et al., 2014; Dawes et al., 2016).
Participatory open data	As an ideal model wherein citizens and governments enter into a dialogue via the OGD initiative, there is active engagement and participation of citizens in the policy-making and policy-implementation stages. Citizens are encouraged to freely contribute toward the existing datasets via the online portal. Furthermore, all grievances of the citizens pertaining to the quality of the datasets are rectified in the prescribed manner. Datasets are qualitatively superior in this model with metadata and permit statistical analysis, interpretation, visualization and mapping.

website quality is linked with user-friendliness, effective website usage and content-related issues and accessibility. Four types of websites have been identified in the study (Table 12.3).

Furthermore, the website quality dimensions have been identified as presented in Table 12.4.

Table 12.5 summarizes the website attributes in terms of the quality parameters.

Table 12.2 Model proposed for evaluating OGD portals (Machova et al., 2018)

Dimension	*Criteria*	*Description*
1. *Open dataset specifications*	a. *Description of dataset*	Portal provides datasets together with their description and how and for what purpose they were collected
	b. *Publisher of dataset*	Portal provides information about organization that published datasets
	c. *Thematic categories and tags*	Portal provides thematic categories of datasets to address the main topics covered. It distinguishes categories (themes) from tags (keywords)
	d. *Release date and up to date*	Portal provides datasets associated with a specific time or period tag, that is, date published, date updated and its frequency
	e. *Machine readable formats*	Portal provides datasets formats that are machine-readable and allow for easy re-use
	f. *Open data license*	Portal provides license information related to the use of the published datasets
	g. *Visualization and analytics tools*	Portal provides visualization and analytics capabilities to gain information about a dataset, e.g., in charts or visualizations in maps
2. *Open dataset feedback*	a. *Documentation and tutorials*	Portal provides high quality of documentation and tutorials to help users in learning how to use the portal
	b. *Forum and contact form*	Portal provides an opportunity to submit feedback on a dataset from the users to providers and forum to discuss and exchange ideas among the users
	c. *User rating and comments*	Portal provides capabilities allowing the collection of user ratings and comments on a dataset
	d. *Social media and sharing*	Portal provides the integration with social media technologies to create a distribution channel for open data and sharing feedback
3. *Open dataset request*	a. *Request form*	Portal provides a form to request or suggest new type or format type of open data
	b. *List of requests*	Portal provides a list of requests that were received from users, including the current state of request processing
	c. *Involvement in the process*	Portal provides capabilities allowing the involvement in the active requests, that is, express interest in the same dataset

Table 12.3 Four types of websites and their attributes

Website	*Main attributes*
The inviting website (InvW)	Website should always make it possible for the user to make the right choice; Users should be able to find information fast as part of effective use; Website development is a continuous process where there is always room for improvement; Website should be lucid and it should be easy to find things; Personalization and speed; Well-developed content and use of pictures; Users should be satisfied and provided with help and support; More use of interactivity and new ways of presenting and visualization of web content; Presenting search functions in order to make it easier for finding relevant information, personalizing web content and making updates and continuous improvements
The intuitive website (IW)	Website appears to have a "logical structure" and this helps in the easy access and free movement across the website; Users are able to easily find information; Regular updating of the website contents
The easy-to-use website (EUW)	Website should be lucid, and it should be easy to find things on the web site; Supports user-satisfaction and efficiency of use; Information presented on the web site must be currently valid' Personalization of web content; Interactivity; Provision of true, updated and trustworthy information; Use of easy language; Providing faster response for users, fast mail response and feedback to users and deliberate responses; Fast response and feedback to users; Presenting content that users are looking for and expecting to find at websites
The simple website (SW)	Accessibility is highly important, together with user adoption and website content; Users should be able to find the services they need; Content should be organized in a way that clearly separates the different parts from one another; Content should be "easy to find" and well "organized" so that users can discover and retrieve relevant information and services; There is data dependability

Table 12.4 Quality aspects and their dimensions

Quality aspects	*Dimensions*
Information quality (captures the website/application content)	Content Dependability Updated
System quality (measuring the desired characteristics of a system or a website)	Accessibility (standard) Effective use Personalization Search functions Speed Uptime User-friendly
Service quality (covers the overall support delivered by the service provider)	Fast mail response Fast response for the users Service quality (easy help)

Table 12.5 Assessment of website types in terms of the attributes

Major attributes	*Simple website (SW)*	*Easy-to-use website (EUW)*	*Intuitive website (IW)*	*Inviting website (InvW)*
Accessibility	✓	✓	✓	✓
Content	X	✓	✓	✓
Deliberate	X	✓	X	X
Dependability	✓	✓	X	✓
Effective use	✓	✓	✓	✓
Fast mail response	X	✓	X	X
Faster response for the users	X	✓	X	X
Interactivity	✓	✓	X	✓
Personalization	X	✓	X	✓
Search functions	X	X	X	✓
Service quality (easy help)	X	✓	X	✓
Speed	✓	X	X	✓
Updated	X	✓	✓	X
Uptime	✓	X	X	X
User-friendly	✓	✓	✓	✓

Evaluation of the National OGD Portal: A Case Study of the UAE's OGD Initiative

UAE occupies the 17th spot out of 137 countries in the Global Competitiveness Report (2017–2018). The country has been adopting economic diversification drives. Given the prospects of OGD initiatives in furthering economic growth, we posit that the OGD initiative of the country has the potential of aiding the country in its economic diversification moves. Apart from furthering economic growth, the OGD initiative should facilitate citizen participation and collaboration in improvising public services and making them more efficient and effective. Datasets published under the aegis of the OGD initiative may be re-used by a diverse set of stakeholders (academic community, scientists, researchers, public/private/non-profit professionals, journalists, software developers, etc.). By re-using the datasets, the stakeholders may derive social and economic value and spearhead innovation. However, to realize the benefits of re-using the datasets, it is important that the datasets published by the government are of good quality and the government is proactive in instituting a robust OGD initiative by contributing toward publishing datasets that are qualitatively and quantitatively superior. Datasets should be complete in themselves and they should be published in a regular manner. This would ensure a sustainable re-use of the datasets by the stakeholders. Therefore, we present an evaluation assessment of the national OGD portal of United Arab Emirates (https://bayanat.ae/) as a case study in order to analyze the quality of the datasets published via the portal. Specifically, the case study addresses the question: "What are the drivers and barriers in re-using the datasets published via the national OGD portal of United Arab Emirates (UAE)?"

The Federal Competitiveness and Statistics Authority (FCSA) is responsible for managing and updating the portal. The portal has a search option wherein keywords may be provided for accessing the datasets. In all, there are 776 datasets covering themes like agriculture (113), labor force (55), education (45), climate (31), and so on. Furthermore, the datasets are provided by organizations like Ministry of Education (40), Ministry of Health (32), Zakat Fund (18), Zayed University (16), Ministry of Energy (15), and so on. Datasets may be filtered using tags like education (53), crop (33), zone (33), federal (31), gender (27), region (27), and so on. Furthermore, datasets are available in two formats: Excel (XLSX) (774) and Excel (XLS) (2). Software applications that may be tapped via open data portal of the country include Aamen (an application for handling employees' complaints about their wages), Absher (an application for achieving the goals and indicators of the national agenda of the country), Dubai Culture (an application for connecting the classical with the modern world), and so on. There is a "Youth Data" platform wherein the innovative efforts of the young generation are being showcased (https://data.youth.gov.ae/en). Visualization of datasets is possible only for select datasets. The government has been arranging hackathons for furthering innovation using open data. The "privacy policy" mentions that any personal information such as name and email address shared by the visitors shall be reserved for confidential purposes only. As per the "terms of use" of the portal, the datasets are available for "free use and reuse for commercial and non-commercial purposes." The portal mentions that the following procedures should not be violated:

- Accessing details that are not intended to be provided to this user or logging into a server or an account that the user is not authorized to access.
- Attempting to conduct any test or survey for finding weakness of any system or network of the UAE government or violating applicable procedures or documenting them without an official permit from the UAE government.
- Attempting to interfere in the provided service on the part of any user, host or network including but not limited to placing a virus on the website, increasing load to or immersing it, sending commercial messages to it or avalanching it with electronic messages or even destroying it.
- Sending unwanted electronic messages to the website including commercials and/or advertisements on services or products or falsifying any dispatch control protocol package address/internet protocol or any part of the address details in any electronic message or sending news messages.
- Using this portal by any means for sending an email, anything of it or on its behalf, by referring to it, or assuming the identity of its name involving offence or libel of the UAE government, website or any person whomsoever, announcing any untrue news or information and ascribing it to the UAE government unrightfully. There is a "contact form" for contributing datasets and soliciting datasets.

Conducting evaluation studies by underlining the features of the national OGD portals has been common till now (See, for instance, Kassen, 2013; 2017; Saxena, 2018). This case study adopts the usability evaluation framework proposed by Machova and her colleagues (2018) as well as the website quality evaluation framework by Sorum et al. (2013). We chose to refer this framework for the present study for three reasons—it is a recent contribution which is relevant for the evaluation of any OGD portal; it is comprehensive and all-encompassing; and it has been derived after a review of literature which shows that it is well-researched and may be subject to further reference for academic research. Table 12.6 summarizes the key

Table 12.6 Major facilitators and hindrances in re-using datasets via the national OGD portal of UAE

Criteria	*Facilitators to re-use datasets via the national OGD portal*	*Hindrances to re-use datasets via the national OGD portal*
1. *Open dataset specifications*		
a. *Description of dataset*	Portal is accessed in two languages—Arabic and English. There is a provision to "search" datasets via the main page. Metadata in the form of data publication date, last revision done, identifiers, spatial/ geographical coverage, authorship and public access level are limited.	Portal provides datasets together with their description and how and for what purpose they were collected. However, real-time data is not being provided via the portal as of now.
b. *Publisher of dataset*	In all, there are 776 datasets covering themes like agriculture (113), labor force (55), education (45), climate (31), and so on. Furthermore, the datasets are provided by organizations like Ministry of Education (40), Ministry of Health (32), Zakat Fund (18), Zayed University (16), Ministry of Energy (15), and so on.	Not all listed publishers have contributed to the portal.
c. *Thematic categories and tags*	Datasets may be filtered using tags like education (53), crop (33), zone (33), federal (31), gender (27), region (27), and so on.	Portal provides thematic categories of datasets to address the main topics covered. It distinguishes categories (themes) from tags (keywords)

(*continued*)

Table 12.6 (continued)

Criteria	*Facilitators to re-use datasets via the national OGD portal*	*Hindrances to re-use datasets via the national OGD portal*
d. *Release date and up to date*	Metadata reveals some information regarding the date of publication of the datasets.	It is important that the metadata be relevant and complete in all aspects.
e. *Machine readable formats*	Datasets are available in two formats: XLSX (774) and XLS (2).	User-friendly formats should be provided for analysis and statistical interpretation.
f. *Open data license*	Data are regarded as a "natural resource" and should be available for all users.	N/A
g. *Visualization and analytics tools*	Datasets may be arranged in an ascending or descending order.	Visualization and mapping tools are required for optimum re-use of the datasets.
2. *Open dataset feedback*		
a. *Documentation and tutorials*	The manual of "Open Data Policy" provides information regarding the OGD initiative of the country and the utility of the same for the different stakeholders.	N/A
b. *Forum and contact form*	There is a "Contact" form where the users/portal visitors may place requests or contribute datasets.	A blog may be provided where the users may interact with each other and lend constructive views for the improvisation of the portal.
c. *User rating and comments*	Datasets may be ranked (five-star rating).	Users may provide "feedback" regarding the website via the "Contact" form.
d. *Social media and sharing*	Social media networks are not provided.	Social networking plugins may be provided (Twitter, Facebook and YouTube) to facilitate the sharing and discussion of datasets.
3. *Open dataset request*		
a. *Request form*	Only the government agencies may place "requests" for publishing the datasets.	Other stakeholders should be involved in the data publishing process.
b. *List of requests*	The provision of providing the list of requests made by the users is missing in the portal.	The portal should provide information regarding the recommendations and contributions of the users to the datasets via the portal.

(*continued*)

Table 12.6 (continued)

Criteria	*Facilitators to re-use datasets via the national OGD portal*	*Hindrances to re-use datasets via the national OGD portal*
c. *Involvement in the process*	Public participation has been encouraged wherein recommendations from the portal visitors are invited for supporting the OGD initiative of the country.	Events like contests for the stakeholders to improvise upon the datasets should be organized. For instance, software apps may be devised by the software professionals. Businessmen may showcase how they harnessed datasets for improvising their services. Implicitly, incentives should be provided for increased involvement of the stakeholder groups.

observations for the national OGD portal of the country in line with the model (Machova et al., 2018). From the table, it is clear that the OGD portal provides opportunities to re-use datasets and the re-use of the datasets by a large cross-section of stakeholders might result in innovation and improvisation of goods and services thereby contributing to economic development of the country. At the same time, there are challenges in re-using the datasets because the datasets need to be updated on a regular basis and a more proactive stance of the government agencies is required to facilitate publishing and interoperability of the datasets.

Table 12.7 summarizes the key observations from the perspective of Sorum et al. (2013). These observations are derived after a careful perusal of the national open data portal of UAE.

Conclusion

The case study sought to present an evaluation of the national OGD portal of the UAE. After a brief scan of the literature on OGD, the case study provided a brief regarding the drivers and barriers to re-use the datasets published via the OGD portal conceding that data re-use by a diverse group of stakeholders is one of the prime rationale for publishing them and making them available for all. The evaluation of the OGD portal was conducted using the recently-proposed framework (Machova et al., 2018)

Table 12.7 Key insights about the portal using the framework of Sorum et al. (2013)

Major attributes	*Major insights derived from the portal analysis*
Accessibility	The national open data portal is well-accessible for use and reference by the relevant stakeholders. Datasets must be made more accessible in terms of their qualitative and quantitative aspects.
Content	The content of the open data portal should be well-maintained and updated on a regular basis.
Deliberate	The national open data portal of the UAE needs to provide real-time information.
Dependability	It is important that the national open data portal provides datasets that are reliable and dependable.
Effective use	The open data portal must provide more user-friendly features for facilitating the re-use of datasets.
Fast mail response	It is important that the mail responses be provided at the right time.
Faster response for the users	The national open data portal must ensure that responses to any queries made by the visitors are being answered promptly.
Interactivity	The national open data portal must facilitate interactivity by incorporating more user-friendly features.
Personalization	The content of the national open data portal must ensure that there is personalization for ensuring that the visitors perceive that the portal is serving their ends well.
Search functions	The website offers a "search" facility so that the specific dataset may be accessed easily depending upon the keyword/s used.
Service quality (easy help)	The website needs to improve on this feature.
Speed	The national open data portal should provide datasets in a speedier manner.
Updated	It should be the priority of the government to publish updated datasets.
Uptime	It would be a better exercise if efforts are made to publish linked datasets on a real-time basis.
User-friendly	The website should provide user-friendly features in order to facilitate a wider range of stakeholders to re-use the datasets.

and the framework proposed by Solem et al. which are quite comprehensive. It may be deduced from this evaluation of the OGD platform that while the OGD initiative of the country is a forthcoming measure of the government for promoting transparency in administration and forging ties with the citizens, the portal needs to be improvised in terms of the quality of the datasets. Datasets should be published on a timely basis and regular updates should be made whenever required. For ensuring increased participation and engagement of the users, incentives should be provided for data re-use. Users' contribution to the datasets should be encouraged and interaction via the platform is also encouraged such that the users may interact among themselves and discuss about the improvisation of the goods and services through data re-use. Datasets lack visualization and mapping tools and this is a bottleneck in analyzing the datasets. Likewise, datasets are not amenable to statistical interpretation on account of the absence of compatible features which may facilitate the re-use of datasets in an optimum manner.

Social and Practical Implications

The case study underlines that the success of any OGD initiative lies in its sustainability. For ensuring the sustainability of the OGD initiative, it is important that the datasets be re-used for creating value out of the same. For instance, datasets may be re-used by a cross-section of stakeholders (citizens, journalists, software experts, academia, entrepreneurs, etc.) for appreciating public administration and providing recommendations for policy-making and improvisation in public service delivery. At the same time, they may tap the datasets for improvisation and innovation of their services. On the other hand, the policy-makers may solicit views from the users regarding the portal and the quality of the datasets published via the same and this would help in improving upon the administrative services. In turn, this would lead to better relationship between the policy-makers and the citizens. It is also important for the government to ensure that a robust ICT infrastructure is in place for sustaining the OGD initiative. Thus, on the one hand, the public officials managing the OGD initiative should be trained with regard to the maintenance and monitoring of the OGD portal and data publishing process; and on the other hand, technological advancement should be upheld for providing real-time linked datasets in near future. This would further improvisation in public service delivery and facilitate citizen participation in the co-creation of services.

Further Research Directions

The case study leaves pointers for further research. For instance, further research is merited to assess the manner in which the OGD initiative is being perceived by the different stakeholder groups. Empirical investigation is required to appreciate the utility of the datasets for different stakeholder groups. Future research may be conducted to analyze the perspectives of the government representatives regarding the factors that impact the data publishing processes. A comparative approach may be undertaken to analyze how lessons may be drawn from the successes of OGD initiatives elsewhere. Research cues may be derived from the significance of data mining and social networks in furthering OGD discussions by the stakeholders. Likewise, it may be interesting to evaluate the significance of Big Data analytics in understanding the efficacy of OGD initiatives (Saxena, 2016). Finally, further research is warranted to analyze the utility of the implementation of OGD initiatives at the regional and local levels of the country.

References

Afful-Dadzie, E., & Afful-Dadzie, A. (2017). Open government data in Africa: A preference elicitation analysis of media practitioners. *Government Information Quarterly, 34*(2), 244–255.

Alexopoulos, C., Loukis, E., & Charalabidis, Y. (2014). A platform for closing the open data feedback loop based on Web 2.0 functionality. *eJournal of eDemocracy & Open Government, 6*(1), 62–68.

Attard, J., Orlandi, F., Scerri, S., & Auer, S. (2015). A systematic review of open government data initiatives. *Government Information Quarterly, 32*(4), 399–418.

Charalabidis, Y., Alexopoulos, C., & Loukis, E. (2016). A taxonomy of open government data research areas and topics. *Journal of Organizational Computing and Electronic Commerce, 26*(1/2), 41–63.

Dawes, S. D., Vidiasova, L., & Parkhimovich, O. (2016). Planning and designing open government data programs: an ecosystem approach. *Government Information Quarterly, 33*(1), 15–27.

Huijboom, N., & Van den Broek, T. A. (2011). Open data: An international comparison of strategies. *European Journal of ePractice, 12*(1), 4–16.

Jaeger, P. T., Bertot, J. C., & Shilton, K. (2012). Information policy and social media: Framing government—Citizen Web 2.0 interactions. In C. G. Reddick & S. K. Aikins (Eds.), *Web 2.0 technologies and democratic governance* (pp. 11–25). New York: Springer.

Janssen, M., Charalabidis, Y., & Zuiderwijk, A. (2012). Benefits, adoption barriers and myths of open data and open government. *Information Systems Management, 29*(4), 258–268.

Jung, K., & Park, H. W. (2015). A semantic (TRIZ) network analysis of South Korea's "Open Public Data" policy. *Government Information Quarterly, 32*(3), 353–358.

Kalampokis, E., Tambouris, E., & Tarabanis, K. (2011a). *Open government data: A stage model.* Paper presented at International Conference on Electronic Government, 28 August–2 September 2011, Springer, Delft, The Netherlands.

Kalampokis, E., Tambouris, E., & Tarabanis, K. (2011b). A classification scheme for open government data: Towards linking decentralised data. *International Journal of Web Engineering and Technology, 6*(3), 266–285.

Kassen, M. (2013). A promising phenomenon of Open Data: A case study of the Chicago Open Data Project. *Government Information Quarterly, 30*(4), 508–513.

Kassen, M. (2017). Open data in Kazakhstan: Incentives, implementation and challenges. *Information Technology & People, 30*(2), 301–323.

Machova, R., Hub, M., & Lnenicka, M. (2018). Usability evaluation of open data portals: Evaluating data discoverability, accessibility, and reusability from a stakeholders' perspective. *Aslib Journal of Information Management.* https://doi.org/10.1108/AJIM-02-2018-0026

Martin, C. (2014). Barriers to the open government data agenda: Taking a multi-level perspective. *Policy & Internet, 6*(3), 217–240.

Mergel, I. (2015). Open collaboration in the public sector: The case of social coding on GitHub. *Government Information Quarterly, 32*(4), 464–472.

Mpinganjira, M. (2015). Use of e-government services: The role of trust. *International Journal of Emerging Markets, 10*(4), 622–633.

Nugroho, R. P., Zuiderwijk, A., Janssen, M., & de Jong, M. (2015). A comparison of national open data policies: Lessons learned. *Transforming Government: People, Process and Policy, 9*(3), 286–308.

Rothenberg, J. (2012). *Case study international benchmark: Towards a better supply and distribution process for open data.* Retrieved from Logica Business Consulting, The Netherlands.

Saxena, S. (2016). Integrating Open and Big Data via 'e-Oman': Prospects and issues. *Contemporary Arab Affairs, 9*(4), 607–621.

Saxena, S. (2017a). Significance of open government data in the GCC countries. *Digital Policy, Regulation and Governance, 19*(3), 251–263.

Saxena, S. (2017b). Prospects of open government data (OGD) in facilitating the economic diversification of GCC region. *Information and Learning Science, 118*(5/6), 214–234.

Saxena, S. (2018). Open government data (OGD) in six Middle East countries: An evaluation of the national open data portals. *Digital Policy, Regulation and Governance, 20*(4), 310–322.

Saxena, S., & Janssen, M. (2017). Examining open government data (OGD) usage in India through the UTAUT framework. *Foresight, 19*(4), 421–436.

Sieber, R. E., & Johnson, P. A. (2015). Civic open data at a crossroads: Dominant models and current challenges. *Government Information Quarterly, 32*(3), 308–315.

Sorum, H., Andersen, K. N., & Clemmensen, T. (2013). Website quality in government: Exploring the webmaster's perception and explanation of website quality. *Transforming Government: People, Process and Policy, 7*(3), 322–341.

Ubaldi, B. (2013). Open government data: Towards empirical analysis of open government data initiatives. *OECD Working Papers on Public Governance*, 22. https://doi.org/10.1787/5k46bj4f03s7-en

Verhulst, S., & Young, A. (2016). *Open Data impact, when demand and supply meet: Key finding of the open data. Impact case studies.* Retrieved from http://odimpact.org/key-findings.html

Wirtz, B. W., & Birkmeyer, S. (2015). Open government: Origin, development, and conceptual perspectives. *International Journal of Public Administration, 38*(5), 381–396.

Wirtz, B. W., Weyerer, J. C., & Rosch, M. (2017). Open government and citizen participation: An empirical analysis of citizen expectancy towards open government data. *International Review of Administrative Sciences.* https://doi.org/10.1177/0020852317719996

Zuiderwijk, A., & Janssen, M. (2014). Open data policies, their implementation and impact: A framework. *Government Information Quarterly, 31*(1), 17–29.

Zuiderwijk, A., Janssen, M., & Dwivedi, Y. K. (2015). Acceptance and use predictors of open data technologies: Drawing upon the unified theory of acceptance and use of technology. *Government Information Quarterly, 32*(4), 429–440.

CHAPTER 13

Artificial Intelligence and Robotics Technology in the Hospitality Industry: Current Applications and Future Trends

Li Yang, Tony L. Henthorne, and Babu George

Introduction

Artificial intelligence (AI) and robotics science have been penetrating into different kinds of businesses and industries in recent years. AI can be defined as the experimental and theoretical study of perceptual and intellectual processes using computers and to make a computer perceive, understand, and act in ways now possible only for humans (Achacoso, 1990). Robots, as one kind of human-like AI, were first introduced into manufacturing industries for product processing in order to replace monotonous human work decades ago (Murauskiene, 2015). For now, the applications of AI and robotics science are extending into other industries, for example, into transportation as autonomous vehicles (Maurer,

L. Yang (✉) • T. L. Henthorne
University of Nevada, Las Vegas, Las Vegas, NV, USA
e-mail: li.yang@univ.nevada.edu; tony.henthorne@unlv.edu

B. George
Fort Hays State University, Hays, KS, USA
e-mail: bpgeorge@fhsu.edu

B. George, J. Paul (eds.), *Digital Transformation in Business and Society*, https://doi.org/10.1007/978-3-030-08277-2_13

Gerdes, Lenz, & Winner, 2016), into the medical industry for diagnoses and surgeries (Kaur, 2012; Mirheydar & Parsons, 2013), into the education industry for teaching or helping special kids (Fridin & Belokopytov, 2014; Timms, 2016), into the supply chain management for organizing warehouse (Min, 2010), and so on.

In the realm of hospitality and tourism, AI and robots have also arrived and they are catching peoples' attention. The world's first robotics hotel, The Henn-na Hotel, was opened in 2015 in Japan, which is fully staffed by robots (Rajesh, 2015). Meanwhile, in the United States, the Wynn Las Vegas announced in December 2016 that they are going to partner with Amazon to equip all of their hotel rooms with an Amazon Echo, a voice activated system (Raz, 2016). Besides the hotel sector, AI and robotics science are finding their way into other sectors of hospitality. In 2015, Eatsa, a futuristic restaurant where robot cubbies serve quinoa bowls, opened in San Francisco, and later opened another restaurant in NYC (Constine, 2015; Kowitt, 2016). A new bar on the Strip of Las Vegas, Tipsy Robot, featuring two robot bar tenders, started their business on June 30th, 2017 (Stapleton, 2017). AI and robotics science not only offers numerous opportunities to hospitality firms to enhance their daily operations, but also ensures hospitality firms provide consistent quality of products and services to their customers (Perez, 2014). The application of AI in the hospitality industry has caught a few, although not a sufficient amount of, scholars' attention (Borràs, Moreno, & Valls, 2014).

This paper will review current literature in the area of the application of AI and robotics science in the hospitality industry. Our examination will focus on the three areas of the application of AI and robotics science. First, it examines the current status and current application of robotics science in human daily life, highlighting their applications in each segment of the hospitality industry; further, it evaluates the impact of AI and robots on three areas: labor and employment, customers, and corporations and communities; lastly, it explores the future potential application of robots and AI in order to see what benefits or risks can be derived from their application in the hospitality industry.

Business Application of Robotics

The Robot Institute of America, a trade association of robot manufacturers and users, defines robots as follows: "a reprogrammable multifunctional manipulator designed to move material, parts, tools, or specialized devices, through variable programed motions for the performance of a variety of tasks." (OTA, 1982, p. 8). Currently, robots are labeled as "intelligent physical devices" (Chen & Hu, 2013, p. 161) which have the capability of autonomy, mobility, and sensory to execute intended duties (International Organization for Standardization, 2012; Tan, Mohan, & Watanabe, 2016). Robotics science is definitely changing the way that humans live and work all over the world, and the application area of AI and robotics science is widening (Veruggio & Operto, 2008). Robots are extending their way out of the factories and into normal life: hospitals, schools, shopping malls, airports, hotels, restaurants, and even our homes. Robots are not just automated machines anymore; they are becoming consumer items (Veruggio & Operto, 2008). In certain scenarios, they are even more than an intelligent tool; they could even be considered a partner or a companion (Louie, McCaoll, & Nejat, 2014)

Types of Robots

The International Organization for Standardization (2012) claims most robots can be categorized into two groups: industrial robots and service robots. The categories are based on the robot's field of application as well as the industrial or non-industrial environments they operate in. Some scholars, further, separate service robots into two groups: professional service robots and personal service robots (Murphy, Hofacker, & Gretzel, 2017).

Industrial Robots

These are usually adopted to perform industrial tasks which can be indicated from the name. This is the initial field where robots were applied in manufacturing and production. According to the International Federation of Robotics, industrial robotics can be classified as either linear robots, Selective Compliance Assembly Robot Arm (SCARA) robots, articulated robots, parallel robots, or cylindrical robots based on the mechanical structure (IFR, 2016a).

Service Robots

"A service robot is a robot which operates semi or fully autonomously to perform services useful to the well-being of humans and equipment, excluding manufacturing operations" (IFR, 2016a). Different from industrial robots, service robots are designed to assist, serve, and interact with humans. International Federation of Robotics (IFR) claims that service robots do not have to have arms, like industrial robots do. According to the definition, an industrial robot could also become a service robot when it is installed in a non-manufacturing environment (IFR, 2016b). Because of the increasing market demand and the different applications, the realm of service robots is expanding year by year, and the variety of service robots is becoming wider. Additionally, their diversity and application is becoming more varied each year. The shape and structure of service robots might change over time as well, but generally service robots can be classified into professional service robots and personal/domestic service robots (Ivanov, Webster, & Berezina, 2017; Murphy et al., 2017).

Professional Robots

From the name, they are deployed by formal organizations and companies. Professional service robots have a much bigger market than domestic/personal robots (IFR statistical department: World Robotics News, 2012). From those statistics, defense, field, logistic, and medical robots are the leading players for professional robots.

Personal/Domestic Robots

Personal/Domestic robots are robots used at home or at other private places by individuals for non-commercial purpose. There are two leading kinds of personal robots. One is cleaning robots, such as floor cleaning, lawn-mowing, and so on. The other is interactive robots, such as butler robots, that are usually equipped with autonomous navigation systems, touchscreens, voice, and face recognition systems. These additional systems can allow them to help or entertain people in public environments or deliver items (Ciupe & Maniu, 2014).

Current Status in Each Hospitality Segment

In the realm of the hotel and tourism industry, AI and robots have also arrived and are attracting great attention and interest from public (Ivanov et al., 2017). The current application of AI and robots in different segments of the hospitality industry, especially service robots will be discussed in this review.

Hotel

Service automation and robotic technologies have made their way into the lodging segment of the hospitality industry, affecting different areas of hotel operations. At the front desk, hotel operators apply self-service kiosks which allow guests to check in and out by themselves in order to reduce waiting time and to improve service (Kim & Qu, 2014). Just like using a kiosk at the airport to get flight tickets, guests can quickly and efficiently check in to their room without human interaction. This is just the very basic application of service automation. In the concierge department, Connie, a robotic concierge agent, was introduced in Hilton Hotels last year, which is partnered with IBM. Connie is powered by AI and can communicate with hotel guests by answering their questions about hotel amenities and services and providing information for local attractions and activities. Because of the AI principle, Connie can also learn from its interactions with guests, and optimize its future answers (Hilton, 2016). In the housekeeping department, a robot called maidbot is created to provide better hotel room cleaning solutions (Maidbot, n.d.). A few hotels, such as Aloft at Cupertino, California, also implement the automated bellboy, a robot that delivers towels and bottled water at the requests of guests (Markoff, 2014).

Combining all these functions, as mentioned earlier in this paper, the world's first robotics hotel, the Henn-na hotel, was opened in 2015 in Japan (Henn-na Hotel, n.d.), which is fully staffed by robots. When guests check into their hotel room, two front desk agents (one human-like robot speaks Japanese, one dinosaur-like robot speaks English) will welcome them; automated trolleys take guest's luggage to their room; and a block-shaped robot does room service as well as delivers beverages and simple snacks.

Food and Beverage

AI and robots are also applied in the food and beverage industry, mostly in restaurant service and food preparation. In some restaurants, the food order is taken by robots (Curtis, 2016). Pizza Hut in Japan hires a humanoid robot named Pepper to take food orders from customers by voice. Pepper is equipped with a voice recognition and AI system, and talks with customers. In addition, Pizza Hut has partnered with MasterCard to program Pepper with a payment app. In other words, Pepper not only takes food orders from customers and sends them to the restaurant kitchen, but also processes orders for customers which saves time for both the restaurant and the customer. As mentioned earlier, a fully automated restaurant, Eatsa, which serves quinoa bowls, opened their first location in San Francisco. Customers place their order through an iPad in the lobby, and their name will show up on one of the transparent LCD screen boxes. In less than 2 mins, customers can retrieve their personalized quinoa bowl out of the box. Similarly, Tipsy Robot, a new bar in Las Vegas, is staffed by two robot bar tenders that mix up and serve drinks (Stapleton, 2017). The guests just need to order their drinks on an iPad, and choose from 18 specialty cocktails or create their own drink. The two robotic bar tenders will get the drink ready for the customers in 60 to 90 seconds. In their spare time, the robots will slice fruit or dance to entertain the customers.

All of these robots utilized in the restaurant mitigate the customer's waiting time and eliminates human waiters and cashiers. Robotics technology definitely helps the food and beverage industry save labor cost and keep food quality consistent.

Meetings and Events

AI and robots are also applied in the meeting and convention industry to provide better service and make the demonstration of meetings easier. A lot of corporations and organizations employ interactive robots at their convention booths or at their trade show (S. P. Events, n.d.). Interactive robots at conventions are usually equipped with a mechanical head and arms, a voice and face recognition system, a message screen or iPad, a CD player, and a brochure holder. Some are also fully mobile so as to attract people's attention or sense people when they pass by. S.P. Events claims that this kind of robot is perfect for promotion events, exhibition stands, tradeshows, or products launched by using their AI tool to communicate with people, introduce products on a screen, or even just to greet people to get their attention. Moreover, this application creates a new robot rental business.

Another significant application of robots in the meeting and events industry are drones. For instance, the organizers of a music festival in South Africa deliver beer to over 150 guests by drone (Daily News, 2013). This application not only provides better service, but also saves the labor cost of delivery for the organizers. Drones are also used during weddings as photographers to take beautiful photos for the new couple.

These examples show that AI and robotics science not only offer numerous opportunities to hospitality firms to enhance efficiency of their daily operations, but also ensure they provide a consistent quality of products and services to their customers.

The Impact of Robots in the Hospitality Industry

Impact on Labor and Employment

Positive Impact
The rapid rise of advanced technology has caused deep uncertainty about the future of the job market (Yakowicz, 2016). There are many kinds of AI and robotics technologies in applied hotels, but there is very little academic research explicitly investigating the impact of robots on the workforce of the hospitality industry. Noone and Coulter (2012) argue that modern AI and robotics science can improve restaurant operations because of the two key features of robotics technologies: sensing for demand prediction and planning for production management. They also claim that robotics should be applied to augment workers' cognitive capability rather than direct labor replacement. Robots can help human staff work more efficiently, rather than replacing their jobs, which will positively impact the business (Pacurar, 2017). From a staff perspective, they also think robots do a great job helping out and making jobs better (Yakowicz, 2016).

Other scholars believe the application of AI in the point of sale system of quick service restaurants can avoid employee theft. The National Restaurant Association (1999) reported that the cost of employee theft for its members totaled over $8.5 billion, or 4 percent, of food sales. Stealing cash from an employer is a significant issue. From that number alone, manager and business runners should not underestimate the lure of quick money and the problem of stealing. Therefore, the most effective way to stop this kind of theft is to apply AI detection systems which aim to maintain solid cash controls for the restaurant (Collins, 2013)

Create New Jobs
Robots are becoming possible alternatives, and they have more of an economic advantage over human beings, in some fields. It is clear that robots are becoming a viable alternative to humans as the economic benefit of using a robot can far outweigh the use of a human. The impact of robots and AI could be dramatic on individuals due to different skillsets required in different industries (OTA, 1982). It is not surprising that labor-intensive and dangerous jobs are replaced by robots or AI. As robots are equipped with cognitive systems, the leading job replaced by robots might be transportation, logistics, customer service, and consumer services. Robots are

replacing humans in some of these jobs in order to reduce the human risk and prevent workers from being exposed to dangerous situations or hazardous environments (Chijindu & Inyiama, 2012; Yakowicz, 2016).

History shows the negative impact of technology on employment has been small enough to be insignificant or undetectable. For example, the wrongly created fear that was aroused in the early 1930s about whether vending machines will replace retail store clerks (Slade, 2007, p. 67). And the truth is people still go to the store. Hence, most experts believe robots or AI can create new jobs rather than replace jobs done by human beings (Veruggio & Operto, 2008; Chijindu & Inyiama, 2012). The production of robots and related technologies will create new jobs, but the amount of jobs created will depend on the growth of the robotics industry and the degree that robots can be automated or self-fixed.

Overall, the literature shows that robotics technology will eliminate some jobs, but create more jobs, thus life will get better in the long run.

Impact on Customers

The application of service robots in the hospitality industry is changing the nature of service, and influencing customers' service experiences and customers' relationships with service providers (Ostrom, Parasuraman, Bowen, Patrício, & Christopher, 2015; Rust & Huang, 2014). Because of this new environment, the interaction between customers and organizations might change significantly. To enhance customer experience, companies will need to optimize robotics technology for the customer's frontline experience, thus, to effectively engage customers and develop relationships. A few scholars are researching these topics. Pan, Okada, Uchiyama, and Suzuki (2015) conducted experiments that tested which style of speech attracted guests in a hotel. According to their experiment data "direct speech was more attractive than indirect speech at the beginning period of interaction between guests and the robot"; thus, they recommend the hotel assistive robots should be designed to provide direct verbal interaction with individual guests in order to engage with guests and to quickly attract their attention.

The Design of Human-like Robots or Machines

Other scholars investigate how the look of robots impacts customers. Mori, MacDorman, and Kageki (2012) suggest if artificial service agents look like a human too much, it would be perceived as creepy; therefore, less warmth for the service. Likewise, Doorn et al. (2017) claim a similar impact will be perceived on the capability dimension of social cognition. In other words, the more the service robots look like a human, it would more likely be conceived as a human and feature lower capabilities. They also note a positive result that because of this verisimilitude, when the human-like robots cause service failures, the customers are more forgivable. Facial gender cues of human-like robots were also examined to investigate their influence on customer behaviors. Eyssel and Hegels' study (2012) shows that a short-haired, male robot is considered as more agentic while a long-haired, female robot is perceived as more communal. This result can inspire the design of service robots and have them change looks depending on what they are programmed to do. Other research is conducted to investigate if human-like features would influence if customers would repeat their service or not (Fan, Wu, & Mattila, 2016). Their results indicate that when customers are alone, they would prefer traditional services, such as talking to human through the phone instead of using touch screens or online services through the phone. This means hospitality firms can avoid adding human-like features to machines in private areas, such as in-room devices. Conversely, adding human-like machines could be practical when used in public, such as a self-service kiosk. Their findings also show that when serving powerful customers, hotels need to consider public image or social presence when using human-like machines. For those hospitality firms that serve more powerful customers in luxury hotels, such as wealthy people, celebrities, and executive corporate staff; these firms need to think of the location and size of the human-like machine to maximize VIP customer satisfaction.

Impact on Corporations and Communities

Increase Production and Reduce Cost

Most literature regarding this subject explores productivity in the context of industrial robotics. Researchers argue that the robotics technology significantly contributes to industrial productivity. The applications of robots to date have extended from large corporations to small businesses; further, AI and robotics technologies will impact small businesses in different areas positively as well, such as predicating demand and planning production (OTA, 1982; Noone & Coulter, 2012; Ivanov et al., 2017). The financial impact of robots in the hospitality firms cannot be ignored, Murphy et al.

(2017) state that industrial robots in the back of the house and service robots in the front of the house both have financial effects.

Cost will be another concern in the context of the application of robots. Many of the newest robots are less expensive than traditional industrial robots. For instance, industrial robots can cost from $100,000 to $500,000 while collaborative robots, which will be adopted in hotels, are priced from $20,000 to $50,000 each (Heires, 2016). A newly built Fairfield Inn & Suites in San Marcos deployed the service robot "Relay" to deliver food, towels, and other items to the guest rooms. Relay costs $2000 per month under a three-year contract with the robot company, Savioke (Hirsh, 2017). The article also mentioned that the application of robots can greatly reduce labor costs for the night shift in hotels and other hospitality segments. The robotics experts believe prices will drop when robots enter into bigger scale manufacturing. In addition, the development of modern batteries and electronics technologies will be lower the cost, advance performance, and broaden the availability of robotics and its related technologies.

The motivation of hotel operators to apply robotics technology is obvious. The trendy application of robots could easily improve their service to some extent, therefore, enhance their reputation and brand image. However, it is doubtful and unclear if hotels can get substantial financial value by implementing robotics technology and reducing labor costs. For example, recall that the Wynn Las Vegas is going to place an Amazon Echo in each their rooms (Raz, 2016). The price of an Amazon Echo is $180, and the Wynn Las Vegas has 2716 rooms for the Wynn location and 2034 rooms for Encore location. After doing a simple math, the total cost will be $855,000 for their initial installment. The Wynn might receive a discount from Amazon, but the future maintenance fee still cannot be underestimated. Yet, Heires (2016) also thinks the application of AI and robots is worth the cost when robots can collect data and map customers' purchasing behavior by directly interacting with them. In this regard, hotels will be able to offer personalization services and achieve higher customer satisfaction in the near future.

In conclusion, some types of the application of robotics science will increase costs while some will reduce costs. Those robots that are applied and on the right track will mostly prove to be valuable (Wells, 2003).

Education and Training

With the growth of robotics technologies, a number of education and training issues are raised by robotics. Some robotics experts also brought up the importance of improving the education system to help people fit into the evolution of the technology; they argued people should focus more on how to extend their knowledge and skills for the age of robots instead of being worried about being replaced by robots (Yakowicz, 2016). Murphy et al. (2017) argued that robotics technologies have impacts on successful tourism and hospitality education. A good start for the robotic world would be to learn about and more understand digital technologies.

Emerging Applications in Each Hospitality Segment

Hotel

Although AI and robots have already entered different hotel departments, the application of robotics is still nascent. The future potential use of robots for the hotel industry is predicted to be broad. It will include, but not be limited to, the laundry department: robots clean and fold sheets and towels; in-room voice-activated systems to help customers control the lighting, AC, and other devices in the room: such as the Amazon Echo mentioned earlier (Raz, 2016). To offer a seamless guest experience, hotel operators will keep chasing better technologies to develop an integrated mobile service ordering system for their guests. This system will eventually enable guest requests to be fulfilled instantly and in no time (Trejos, 2015). Hotels are going to become fully service automated through self-service devices and robotic technologies in the future.

Food and Beverage

For the future application of robotics technology in the food and beverage industry, Ivanov et al. (2017) stated that if AI can offer more advanced technologies, restaurants will also apply those service automation devices or technologies to pursue a fully automated business model similar to the Henn-na hotel, except a restaurant version. They also claim the must-have elements in the future for automated restaurants. These must-haves should contain both front of the house and back of the house automation: at the front of the house, customers can order food through self-service devices or with a robot waiter; at the back of the house, robot chefs will cook the dishes and deliver the dishes through automation devices, such as conveyors in the sushi bar, or with a robot busser to the customers.

In the quick service restaurant segment, other scholars believe robotic technology or robotic management systems will provide small restaurants opportunities to compete with large chain restaurants, because of their scalability. In other words, although large chain restaurants will have the procurement power and brand equity as competitive advantages, in the scenario of applying affordable robotics management, small businesses will be productive and maintain high efficient daily operations. Robotic management would become the strength of small quick service restaurants (Noone & Coulter, 2012).

Meetings and Events

As mentioned previously, the current application of robots in the meetings industry is mainly to amuse attendees at the meetings. Base on this application, a future potential use of robotics science will combine with other technology, for example, telepresense (MantaroBot, n.d.). Telepresence is an innovative technology for the meeting industry, which allows people who are far away or not available to attend meetings and conventions through the MantaroBot. The key idea of this business model is to enable people to use robots as a physical presence indicator to attend meetings or events (Ivanov et al., 2017). MantaroBots will have screens, cameras, and be mobile like other service robots; in addition, they will have a video phone-call feature.

This technology will be very revolutionary for the meeting industry, and offer the opportunity for more people to attend meetings and conventions virtually without transportation costs, and to still have interaction with others even though they could not physically attend the meetings. A similar idea is to use virtual reality to attend meetings (Revvr Studios, 2017).

The Risks Posed by Developments in Robotics

After forecasting a potential use for robotics technology in the future, it is necessary to explore the possible factors that will influence the development. Kuo, Chen, and Tseng (2017) used expert panel research methods and concluded six factors that will influence the development of robots in the hospitality industry. These six factors are government support, capability for market development, future development of the robotics industry, capability for technology development, capability for raising money, and capability for talent development. Some research also investigates other factors that could influence service robots. Gorle and Clive (2013) noted that a country with a high lifespan and low birth rate will more likely have an increasing demand for service robots, such as Japan.

Other studies investigate how to make the application of robotics technology successful in the hospitality industry. For example, strategic positions of the hotels in a market could impact the success of using robots (Melian-Gonzalez & Bulchand-Gidumal, 2016). Improving customer satisfaction and developing into new markets (Baker & Magnini, 2018), highlighted with the practical execution of the marketing campaigns, are also important elements to the success of advanced services with robots or AI (Nieves & Diaz-Meneses, 2016).

When hospitality companies enjoy the convenience and consistent service from robots, some of them also begin to be concerned about the risk and safety of adopting AI and robots, or possibly even "fire" robots due to certain defects (Mccrum, 2016). Some experts mention this concern is reasonable because of the self-learning skills and big data capabilities of the robots, especially those of the interactive robots (Heires, 2016). They also argued robots are originally safe, but if they mingle with other technologies or platforms that could modify or add other programs into the robots, the robots might not be safe anymore. So for now, the experts think the biggest risk related to robot industry is cybersecurity.

Conclusion

Artificially intelligence (AI) and robotics technology have already penetrated into different segments in the hospitality industry. As a service industry, the majority of robots used in the hospitality industry are service robots. This review mainly discusses the application of professional service robots in the hospitality industry. In hotels, robots are utilized in the front office, housekeeping, and the concierge department. In the food and beverage industry, robots could be servers, chefs, bartenders, and bussers. During meetings and events, robots aim to amuse people to promote products.

With the growth of the robotics industry, the impacts of AI and robots are more and more significant. The AI system can augment staff capability, thus, improve daily operations. Additionally, a robotics management system can eliminate internal staff theft. Although robots might replace a few human's jobs, they will still create more jobs in the labor market. AI and robotics has a positive impact on labor and employment. The research in regard to how robots impact customers is mainly about how looks and interactions between robots and human influences customer behavior, especially with human-like robots. This academic area can help design future robots. AI and robots can have a financial impact on companies, increase their productivity, and reduce the labor costs.

The hotel industry wants to reap from these developments and aims to attain full automation. For meetings and events, robots will combine with other technologies in order to create a more innovative application. Telepresence, which can allow human beings to interactive with each other in a meeting without physical attendance, is a particularly interesting application.

Hotels or other hospitality segments invest a lot of money into AI and robots, to make it foolproof. The biggest risk of utilizing robots in the hospitality industry is cybersecurity, because most of the robots need to connect to the internet. To avoid the risk, they have to make sure that all of these new systems are safe and that hackers cannot recode them.

There is still not a lot of research into the application of robotics science or AI in the context of the hospitality industry. Future research may benefit by investigating the following areas: in what locations should robots be placed in a hotel, restaurant, or even at a meeting in order to influence customers' perceptions of robots; what characteristics (e.g. age, nationality, race) should robots have to improve the service experience? Robots have already entered our lives in various guises. Humans have a lot to do to optimize the application of robots and to make our lives even better.

References

Achacoso, T. B. (1990). Artificial ethology and computational neuroethology: A scientific discipline and its subset by sharpening and extending the definition of artificial intelligence. *Perspectives in Biology and Medicine, 33*(3), 379–390.

Baker, M. A., & Magnini, V. P. (2018). The evolution of services marketing, hospitality marketing and building the constituency model for hospitality marketing. *International Journal of Contemporary Hospitality Management, 28(8)*, 1510–1534.

Borràs, J., Moreno, A., & Valls, A. (2014). Intelligent tourism recommender systems: A survey. *Expert Systems with Applications, 41*(16), 7370–7389.

Chen, Y., & Hu, H. (2013). Internet of intelligent things and robot as a service. *Simulation Modelling Practice and Theory, 34*, 159–171.

Chijindu, I., & Inyiama, H. C. (2012). Social implications of robots—An overview. *International Journal of Physical Sciences, 7*(8), 1270–1275.

Ciupe, V., & Maniu, I. (2014). New trends in service robotics. *Mechanisms and Machine Science, 16*, 57–75.

Collins, G. (2013). Safeguarding Restaurants from point-of-sale fraud: An evaluation of a novel theft deterrent application using artificial intelligence. *Journal of Hotel Business Management, 2*(105). https://doi.org/10.4172/2169-0286.1000105

Constine, J. (2015, August 31). Eatsa, A futuristic restaurant where robot cubbies serve quinoa. Retrieved August 8, 2017, from https://techcrunch.com/2015/08/31/eatsa/

Curtis, S. (2016). Pizza Hut hires ROBOT waiters to take orders and process payments at its fast-food restaurants. *Mirror*. Retrieved August 8, 2017, from http://www.mirror.co.uk/tech/pizza-hut-hires-robot-waiters-8045172

Daily News. (2013). Drone delivers beers—not bombs—at South Africa music festival. Retrieved August 8, 2017, from http://www.nydailynews.com/life-style/eats/drone-drops-beers-not-bombs-south-africa-article-1.1422617

Doorn, J., Mende, M., Noble, S. M., Hulland, J., Ostrom, A., Grewal, D., & Petersen, J. A. (2017). Domo arigato Mr. Roboto: Emergence of automated social presence in organizational frontlines and customers' service experiences. *Journal of Service Research, 20*(1), 43–58.

Eyssel, F., & Hegel, F. (2012). (S)he's got the look: Gender stereotyping of robots. *Journal of Applied Social Psychology, 42*(9), 2213–2230.

Fan, A., Wu, L., & Mattila, S. (2016). Does anthropomorphism influence customers' switching intentions in the self-service technology failure context? *Journal of Services Marketing, 30*(7), 713–723.

Fridin, M., & Belokopytov, M. (2014). Acceptance of socially assistive humanoid robot by preschool and elementary school teachers. *Computers in Human Behavior, 33*, 23–31.

Gorle, P., & Clive, A. (2013). Positive impact of industrial robots on employment. Retrieved August 8, 2017, from www. ifr.org/uploads/media/Update_Study_Robot_creates_Jobs_2013.pdf

Heires, K. (2016, September 1). Rise of the robots. Retrieved August 8, 2017, from http://www.rmmagazine.com/2016/09/01/rise-of-the-robots/

Henn-na Hotel. (n.d.). Retrieved August 8, 2017, from http://www.h-n-h.jp/en/

Hilton. (2016). Hilton and IBM pilot "Connie," the world's first Watson-enabled hotel concierge. Retrieved August 8, 2017, from http://news.hiltonworldwide.com/index.cfm/news/hilton-and-ibm-pilot-connie-the-worlds-first-watsonenabled-hotel-concierge

Hirsh, L. (2017, May 13). Hotels make room for robots on staff: Bots could reduce labor costs. Retrieved from http://sdbj.com/news/2017/may/18/hotels-make-room-robots-staff/

International Federation of Robotics (IFR). (2016a). World robotics industrial robots. Executive summary. Retrieved August 8, 2017, from http://www.ifr.org/fileadmin/user_upload/downloads/World_Robotics/2016/Executive_Summary_WR_Industrial_Robots_2016.pdf

International Federation of Robotics (IFR). (2016b). World robotics service robots. Executive summary. Retrieved August 8, 2017, from https://ifr.org/img/office/Industrial_Robots_2016_Chapter_1_2.pdf

International Organization for Standardization. (2012). ISO 8373:2012(en) Robots and robotic devices—Vocabulary. Retrieved August 8, 2017, from https://www.iso.org/obp/ui/#iso:std:iso:8373:ed-2:v1:en:term:2.2

Ivanov, S., Webster, C., & Berezina, K. (2017). *Adoption of robots and service automation by tourism and hospitality companies*. Paper presented at the INVTUR Conference, 17–19 May 2017, Aveiro, Portugal.

Kaur, S. (2012). How medical robots are going to affect our lives. *IETE Technical Review, 29*(3), 184–187.

Kim, M., & Qu, H. (2014). Travelers' behavioral intention toward hotel self-service kiosks usage. *International Journal of Contemporary Hospitality Management, 26*(2), 225–245.

Kowitt, B. (2016, December 13). San Francisco's favorite quinoa spot is taking on New York. Retrieved August 8, 2017, from http://fortune.com/2016/12/13/eatsa-new-york-location-quinoa/

Kuo, C., Chen, L., & Tseng, C. (2017). Investigating an innovative service with hospitality robots. *International Journal of Contemporary Hospitality Management, 29*(5), 1305–1321. https://doi.org/10.1108/IJCHM-08-2015-0414

Louie, W.-Y. G., McCaoll, D., & Nejat, G. (2014). Acceptance and attitudes toward a human-like socially assistive robot by older adults. *Assistive Technology, 26*(3), 140–150.

Maidbot. (n.d.). Retrieved August 8, 2017, from http://www.maidbot.co

MantaroBo. (n.d.). Retrieved on August 8, 2017, from http://www.mantarobot.com

Markoff, J. (2014, August 11). "Beep," says the bellhop: Aloft hotel to begin testing 'botlr,' a robotic bellhop. Retrieved August 8, 2017, from https://www.nytimes.com/2014/08/12/technology/hotel-to-begin-testing-botlr-a-robotic-bellhop.html

Maurer, M., Gerdes, J. C., Lenz, B., & Winner, H. (Eds.). (2016). *Autonomous driving: Technical, legal and social aspects*. Berlin, Heidelberg: Springer Open.

Melian-Gonzalez, S., & Bulchand-Gidmumal, J. (2016). A model that connects information technology and hotel performance. *Tourism Management, 53*(C), 30–37.

Mccrum, K. (2016). Robot waiters fired after spilling drinks and food during restaurant service. Retrieved August 8, 2017, from http://www.mirror.co.uk/tech/robot-waiters-fired-after-spilling-7715093

Min, H. (2010). Artificial intelligence in supply chain management: Theory and applications. *International Journal of Logistics, 13*(1), 13–39.

Mirheydar, H. S., & Parsons, J. K. (2013). Diffusion of robotics into clinical practice in the United States: Process, patient safety, learning curves, and the public health. *World Journal of Urology, 31*(3), 455–461.

Mori, M., MacDorman, K., & Kageki, N. (2012). The uncanny valley. *Robotics & Automation Magazine, IEEE, 19*(2), 98–100.

Murauskiene, N. (2015, May 12). Artificial intelligence is here to manage ATMs. Retrieved August 8, 2017, from https://www.linkedin.com/pulse/artificial-intelligence-here-manage-atms-neringa-murauskiene

Murphy, J., Hofacker, C., & Gretzel, U. (2017). Dawning of the age of robots in hospitality and tourism: Challenges for teaching and research. *European Journal of Tourism Research, 15*, 104–111.

National Restaurant Association. (1999). Avoiding an inside job. *Bread & Butter Newsletter*.

Nieves, J. N., & Diaz-Meneses, G. (2016). Antecendents and outcomes of marketing innovations. *International Journal of Contemporary Hospitality Management, 28*(8), 1554–1576.

Noone, B., & Coulter, R. (2012). Applying modern robotics technologies to demand prediction and production management in the quik-service restaurant sector. *Cornell Hospitality Quarterly, 53*(2), 122–133.

Office of Technology Assessment. (1982). *Exploratory workshop on the social impacts of robotics: Summary and issues*. Washington, DC: Congress of the United States.

Ostrom, A. L., Parasuraman, A., Bowen, D. E., Patrício, L., & Christopher, A. V. (2015). Service research priorities in a rapidly changing context. *Journal of Service Research, 18*(2), 127–159.

Pacurar, A. (2017, January 20). Can artificial intelligence help the hotel industry? Retrieved August 8, 2017, from https://www.cpexecutive.com/post/can-artificial-intelligence-help-the-hotel-industry/

Pan, Y., Okada, H., Uchiyama, T., & Suzuki, K. (2015). On the reaction to robot's speech in a hotel public space. *International Journal of Society Robotics, 7*, 911–920.

Perez, S. (2014). NoWait, The app that lets you join a restaurant wait list from your phone, goes nationwide. Retrieved from https://techcrunch.com/2014/01/30/nowait-the-app-that-lets-you-join-a-restaurant-wait-list-from-your-phone-goes-nationwide/

Rajesh, M. (2015). Inside Japan's first robot-staffed hotel. *The Guardian*. Retrieved August 8, 2017, from https://www.theguardian.com/travel/2015/aug/14/japan-henn-na-hotel-staffed-by-Robots

Raz, N. (2016, December 15). Amazon echo a 'game-changer' for Wynn Las Vegas. *Las Vegas Review-Journal*. Retrieved August 08, 2017, from https://

www.reviewjournal.com/business/casinos-gaming/amazon-echo-a-game-changer-for-wynn-las-vegas/

Rust, R. T., & Huang, M.-H. (2014). The service revolution and the transformation of marketing science. *Marketing Science, 33*(2), 206–221.

Slade, G. (2007). *Made to break: Technology and obsolescence in America*. Boston, MA: Harvard University Press.

S. P. Events. (n.d.). Fairs and promotions. Retrieved August 8, 2017, from http://www.spevents.co.uk/our-events/event-selector/exhibition_stand_attractions/interactive_robot_hire.html

Stapleton, B, (2017, June 30). Go on a Bender at the new Tipsy Robot. Retrieved August 8, 2017, from https://vegas.eater.com/2017/6/30/15898204/tipsy-robot-now-open

Revvr Studios. (2017, February 17). Where are the virtual conventions? Retrieved August 8, 2017, from http://www.revvrstudios.com/where-are-the-virtual-conventions

Tan, N., Mohan, R. E., & Watanabe, A. (2016). Toward a framework for robot-inclusive environments. *Automation in Construction, 69*, 68–78.

Timms, M. J. (2016). Letting artificial intelligence in education out of the box: Educational cobots and smart classrooms. *International Journal of Artificial Intelligence in Education, 26*(2), 701–712.

Trejos, N. (2015). Marriott to hotel guests: We're app your service. *USA Today*. Retrieved August 8, 2017, from http://www.usatoday.com/story/travel/2015/05/13/marriott-hotels-mobile-requests-two-way-chat/27255025/

Veruggio, G., & Operto, F. (2008). Roboethics: Social and ethical implications of robotics. In *Springer Handbook of Robotics*, Siciliano, B., & Khatib, O. (Eds.). Heidelberg: Springer. 1499-1524.

Wells, P. N. T. (2003). Can medical robots reduce health care costs? *Industrial Robot: An International Journal, 30*(1). Retrieved from https://doi-org.ezproxy.library.unlv.edu/10.1108/ir.2003.04930aaa.002

World Robotics News. (2012, August 28). Professional service robots—Continued increase. Retrieved from https://ifr.org/ifr-pressreleases/news/professional-service-robots-continued-increase

Yakowicz, W., & Inc. (2016, October 17). Robots might take your job—but you'll get a better one. *Chicago Tribune*. Retrieved from http://ezproxy.library.unlv.edu/login?url=https://search.proquest.com/docview/1829440782?accountid=3611

CHAPTER 14

Risk Management in the Digital Era: The Case of Nigerian Banks

Tankiso Moloi and Oluwamayowa Olalekan Iredele

Introduction

Moloi (2018) argued that in the context of the public sector, more emphasis tends to be biased to processes and systems, discounting the role of people, which renders the enterprise risk management program unsuccessful. Patterson (2015) concurs with this point as he points out that the enterprise risk management program would only be effective if it is supported by people, processes and technologies. People, processes and systems are all recipes for a successful risk management program. This however requires that the organization has reliable data to base their analysis and decision on.

One of the main pitfalls of the risk management programs in organizations has been a lack of reliable information and data to base decisions on. Patterson (2015) points to two factors that result in unreliable data from the risk management perspective and that is risk data is often scattered across the organization and not shared across business unit, in essence, the culture of silos is an impediment. Further, Patterson points to the fact that

T. Moloi (✉) • O. O. Iredele
University of Johannesburg, Johannesburg, South Africa
e-mail: smoloi@uj.ac.za; oiredele@uj.ac.za

B. George, J. Paul (eds.), *Digital Transformation in Business and Society*, https://doi.org/10.1007/978-3-030-08277-2_14

many risk management functions lack the tools they need to capture and use risk information more effectively. What this effectively means is that it could be expected that there could be opportunities that the organization fails to capture.

By utilizing technology, Ernst and Young (2014) assert that we could begin to see the reduction in missed opportunities and realized risks. This is due to the fact that technology facilitates “improved insight, efficiency and effectiveness in risk management activities, thus freeing up time for risk management resources to spend on risk identification, assessment and control of future risks”. Knopjes (2017) supports this view by stating that organizations could take advantage of emerging technologies to integrate different systems, collect and analyze massive volumes of data from an unlimited number of sources across multiple locations. Through this process, organizations will achieve enhanced operational processes, improve timeliness of reporting and utilize data to drive preventative actions resulting in them being miles head of potential risks (Knopjes, 2017).

Due to the nature of their business, which includes being the repository of liquidity, the core payment mechanism and the principal source of finance to at least a large part of the economy, banks assume a variety of risks (Moloi, 2014; Liu, 2004). The traditional finance theory has limited these risks into major balance sheet risks, namely, liquidity risk, interest rate risk, capital risk and credit risk (Hempel, Coleman, & Simonson, 1990). Traditional finance theory had not extended itself to think about the manner in which these risks and many other facing banks should be governed. As can be noted, there had not been a realization that technology would rise and lead to the circumstance where banks are dominantly relying on it for almost all their activities.

Moloi (2014) argues that banks are important to the economy as they lubricate the financial system, serve as the point of connection between various sectors of the economy and reassure a significant level of specialization, competence, economies of scale, and settings conducive to execution of different economic policies of the government (Sanusi, 2011) and as such, there is a need to identify, understand and monitor the circumstances, which could result in the banking systems’ vulnerability (Moloi, 2014). Akande (2016) and Tennant and Tracey (2014) concur with this and they point out that banks are involved in financial intermediation by sourcing for savings at a rate that will entice depositors and lend funds to creditors at an affordable rate to maximize profits in a competitive environment.

The manner in which these institutions function is even more critical in developing nations such as Nigeria as the country battles to attract foreign direct investments for infrastructure and other needs. The Nigerian banking sector has faced a tumultuous period in recent times which has led to the restructuring and consolidation exercise (Aransiola, 2013). The reform led to the reduction in the number of banks from 89 to 25.

It had been thought that the restructuring and consolidation exercise would lead to stability, however; this situation was short lived as the banks got affected by another financial crisis that resulted in the survival of only 24 banks (Akande, 2016). The challenges that lead to the collapse of these institutions were attributable to risk management. As of 2018, the Central Bank of Nigeria has indicated that there are only 22 commercial banks that are healthy and are standing in Nigeria (CBN, 2019).

As the institution battled the credibility question and lack of confidence, the CBN responded by instituting the CBN code of good governance in order to give guidance to the Nigerian banks on the minimum corporate governance requirements. The central bank indicated that this was developed with the primary purpose of promoting a transparent and efficient banking system that will engender the rule of law and encourage division of responsibilities in a professional and objective manner (CBN, 2014). Adeoye and Amupitan (2014) advise that the concept of good governance in banking industry would need to be stretched in order to empirically include total quality management, which has six performance areas, namely, capital adequacy, assets quality, management, earnings, liquidity and sensitivity risk. The degree of adherence to these parameters determines the quality rating of the banks.

Given the importance of banks in the economy, the recent challenges faced by the banking sector in Nigeria, we are of the view that the evolutions in computing and risk technology which has led to the developments in technologies that could be used to exploit big data, conduct complex analytics, these technical advancements offer the Nigerian banks and authorities such as the CBN with better abilities for enhancing risk management effectiveness, thereby reducing vulnerabilities and exploiting opportunities that could arise.

This study therefore investigated whether or not the Nigerian banking sector was exploiting technology in order to reduce missed opportunities and realized risks, whether technology has been exploited in order to integrate different systems, collect and analyze massive volumes of data from an unlimited number of sources across multiple locations. Essentially, our research question was answered through the use of content analysis of integrated reports of selected Nigerian banks where the focus is on the manner and the way in which risks have been captured. We content-analyzed this section of the integrated report in order to determine whether Nigerian banks indicated in these reports have deployed technology or not and whether or not advanced technology would have been deployed in order to reduce missed opportunities and realized risks.

The remainder of the chapter is organized as follows: Section "A Brief Overview on Corporate Governance" provides a brief overview on corporate governance, Section "A Brief Overview on Recent Corporate Governance Challenges in Nigerian Banks" briefly provides an overview on recent corporate governance challenges in the Nigerian banking sector, Section "Code of Corporate Governance for Banks in Nigeria" discusses the manner in which the codes of good practices including the CBN code are guiding banking firms on the governance of risk and whether or not they do highlight the role of technology, Section "Risk Management in the Digital Age" discusses risk management in the era of digital infrastructure and Section "Empirical Analysis of the Current State of Risk Management of Banks in Nigeria" presents the findings relating to the use of technology and technological infrastructure in risk management.

A Brief Overview on Corporate Governance

The financial crisis around the world and the consequent collapse of major corporate institutions in both developed and developing economies has brought to the fore the issue of corporate governance. Today, corporate governance has attracted considerable attention of policy makers and academic researchers across the globe. The emphasis is on the need for the practice of good governance both at the public and private enterprises and this is due to the economic primacy of publicly quoted firms in most national economies. Corporate governance is increasingly understood among policy makers as a value enhancing strategy in a competitive environment and there is a growing consensus globally that corporate governance has a positive link to national growth and development (Akinkoye & Olasanmi, 2014).

According to Alushna (2017), codes of best practice represent the self-regulation of listed companies and provide a range of recommendations (Larcker & Tayan, 2011; Mallin, 2004; OECD, 2004, 2015; Tricker, 2012) addressing:

- Rights of shareholders, stressing the equal treatment of shareholders holding shares of the same class
- The procedures and rules shaping the functioning of the annual shareholder meeting (ASM) and measures empowering shareholders motivating them to active participation in ASM
- Responsibilities of executives who are accountable to shareholders and stakeholders
- The procedures and rules shaping the functioning of the board
- Transparency standards which describe the scope and content of information policy (company operation and strategy, financial situation, ownership structure, composition, structure and procedures of the board, company by-laws and regulations, executive compensation)

The Cadbury report was the first step where corporate governance code had required that companies follow the "comply or explain" principles, which means that listed firms had to comply with the code guidelines. Those companies that could not comply were expected to report the non-compliance and provide explanation for the non-compliance with the possible measures to be taken for improvement (Cadbury Report, 1992).

Today, pushing for higher governance standards has become a regular campaign with the participation of an increasing number of parties such as academics, media, regulatory authorities, corporations, institutional investors and international organizations shareholders right watchdog amongst others. Corporate failures have necessitated considerable interests on empirical research on the effectiveness of various corporate governance institutions and mechanisms. We note that particular attention has understandably been drawn to addressing and researching the underlying issues and factors that led to the financial and corporate crisis that characterized both the developed and developing economies. Many nations today have taken numerous initiatives such as the introduction of code of best practice, new listing/disclosure rule, mandatory training for board directors and enforcement of code of governance amongst other things; as measures to address and enhance the issue of corporate governance practice. International organizations such as the Basel Committee on Banking Supervision, International Monetary Fund (IMF) and Organisation of Economic Co-operation and Development (OECD) have become major proponents of governance (Moloi, 2009, 2014). In this regard, Akinkoye and Olasanmi (2014) point out that IMF, for instance, demands that governance improvement should be included in its debt relief program. These efforts to ensure good corporate governance practice had also been extended to the banking industry (Moloi, 2016a, b).

A Brief Overview on Recent Corporate Governance Challenges in Nigerian Banks

In the Nigerian banking context, corporate governance has recently received increased attention because of high-profile scandals involving abuse of corporate power, and in some cases, alleged criminal activities by those entrusted with governance. The abuse of corporate power and criminal activities has directly been attributable to weak corporate governance in banks. Recently, the Governor of the Central Bank of Nigeria, Godwin Emefiele went as far as indicating that that corporate governance practices in the Nigerian banking sector leaves much to be desired. His main concern in this regard appears to be the fact that these fraudulent incidents were capable of undermining financial stability by heightening vulnerability of financial institutions to external shocks (The Punch, 2017).

In the banking sector good corporate governance practices are regarded as important in reducing risk for investors, attracting investment capital and improving the performance of companies. The series of widely publicized cases of accounting improprieties recorded in the Nigerian banking industry in 2009, for instance cases involving Oceanic Bank, Intercontinental Bank, Union Bank, Afri Bank, Fin Bank and Spring Bank were related to the lack of vigilant oversight functions by the boards of directors. Accordingly, boards in these banks appear to relinquish governance to corporate managers who then pursued their own self-interests leading to the reported improprieties (Uwuigbe, 2011).

Due to the fact that banks have an important role in the intermediation process, any disturbance in this sector will likely affect the economy. Monnin and Jokipii (2010) posit that modern financial system contributes to economic development and the improvement in living standards by providing various services to the rest of the economy. These include clearing and settlement systems to facilitate trade, channeling financial resources between savers and borrowers and various products to deal with risk and uncertainty. The ability of the bank therefore to manage risk being a major threat to its going concern is crucial, not only for itself but for various stakeholders associated with its business.

Code of Corporate Governance for Banks in Nigeria

Following the conclusion of the consolidation program in 2005, a Code of Corporate Governance for Banks in Nigeria was issued to the banking industry. The code which became effective in April 2006 was designed to enhance corporate governance practices within the banking industry in view of the fact that governance mechanisms in banks was notably weak and board members of financial institutions were unaware of their statutory and fiduciary responsibilities.

The update on the code of governance became necessary in view of the ambiguity in the code and its conflict with the provisions of Companies and Allied Matters Act (CAMA) 1990 as well as the need to align the 2005 code with contemporary developments and international best practices, hence the updated 2014 edition. With its promulgation, there was an expectation that the revised code would provide clear guidelines on all aspects of governance and that it would enhance Corporate Governance practices for banks in Nigeria.

The 2014 code covers seven key aspects of corporate governance as it affects the banking sector in Nigeria, namely, board and management, shareholders, right of other stakeholders, Disclosure in annual reports, risk management, ethics and professionalism and conflict of interest and finally, sanctions. The board has a special role in carrying out an oversight of functions and the effectiveness with which the boards of financial institutions discharge this responsibility is critical to the bank survival. The code requires that boards be free to drive their institutions forward, but exercise that freedom within a framework of transparency and effective accountability.

For the purpose of this study, we indicate that while other aspects of the code are relevant to the going concern of banks, risk management is central in view of the nature of the risks that banks face and the fact that they are in custody of other people's money (Moloi, 2014, 2016a). The CBN code makes it the responsibility of the board of a bank to ensure that policies, procedures and controls are put in place to manage the various types of risks with which it is faced. Given the multifaceted nature of risk, the complexities surrounding today's organizations, particularly banks, governing risks in an institution like a bank would be complex. The speed in which technology is emerging provides an opportunity to the boards to insist to managements of these complex organizations that technology that integrates different systems and collects and analyzes massive volumes of data from an unlimited number of sources across multiple locations is a step forward toward the enhanced operational processes, an improvement in timeliness of reporting and utilization of data for the purpose of driving preventative actions which would result in organizations being miles ahead of potential risks (Knopjes, 2017).

It is the purpose of this study to investigate whether or not the Nigerian banking sector is exploiting technology in order to reduce missed opportunities and realized risks, and further, how technology has been exploited in order to integrate different systems and collect and analyze massive volumes of data from an unlimited number of sources across multiple locations. The objectives of this study are achieved through content-analyzing integrated reports of selected Nigerian banks where the focus is on the manner and the way in which risks have been captured. We have content-analyzed this section of the integrated report in order to determine whether or not Nigerian banks indicate that they utilize technology in risk management and whether these reports mention advance risk management technology and have deployed the same in order to reduce missed opportunities and realized risks.

Risk Management in the Digital Age

The CBN Code of Corporate Governance for Banks and Discount houses provides that

> *[e]very bank shall have a risk management framework specifying the governance architecture, policies, procedures and processes for the identification, measurement, monitoring and control of the risks inherent in its operations. The Board is responsible for the bank's policies on risk oversight and management and shall satisfy itself that management has developed and implemented a sound system of risk management and internal control.*

We live in a rapidly changing world. There is no time in history when virtually every aspect of human life has been affected by the rapid change brought about through information technology (Harari, 2018). According to Marwala and Hurwitz (2017), the world has witnessed four phases of revolution in human history. These revolutions are described below:

- The first industrial revolution brought mechanical innovations with the development of steam engine which was key to the then industrial revolution;
- The second industrial revolution which started in the second half of the nineteenth century brought the oil-powered internal combustion engine and electrical communication. Major technological advances during this period included the telephone, light bulbs and phonograph (Marwala & Hurwitz, 2017);

- The third industrial revolution or digital revolution which came in the 1980s brought computerization, that is, mainframe computers, personal computers and the internet, and the information and communication technology (ICT) available today. This has been a period of advancement of technology from analog electronic and mechanical devices to the digital technology (Agrawal, Gans, & Goldfarb, 2018; Marwala & Hurwitz, 2017); and
- The fourth industrial revolution which has arrived at the beginning of the twenty-first century is the advent of cyber-physical systems representing new ways in which technology becomes embedded within societies, that is, business, government, civil society, and so on, and the human body; it is driven by the rapid convergence of advanced technologies across the biological, physical and digital worlds. It is marked by emerging technology breakthroughs in a number of fields, including robotics, artificial intelligence, biotechnology, and so on. All of these revolutions came with unique impacts on every aspect of human lives including business (Harari, 2018; Agrawal et al., 2018; Marwala & Hurwitz, 2017).

The banking sector is undergoing its own revolution as well with significant implications for risk management (Moloi, 2014; Moloi, 2016a). Risk management is very significant to the operations of any business entity due to serious consequences that the occurrence of risk portends (Moloi, 2016b). It implies that for a business organization to be rest assured of the achievement of its objectives besides survival and growth, risk management becomes imperative (Ayodele & Alabi, 2014). Risk is the exposure to loss arising from the variation between the expected and actual outcomes of investment activities (Nzotta, 2002; Owualla, 2000). Therefore, in a broad term, risk management can be related to a mechanism which embraces planning, organizing and controlling resources and operational activities of business for effective reduction or elimination of risk or the adverse effects of risks (Ayodele & Alabi, 2014). Management of risks is thus imperative for banks given the nature of their business.

According to Ayodele and Alabi (2014), Onyekwelu and Onyeka (2014) as well as Moloi (2014), some of the common risks that are managed by banks include:

- Credit (default) risk which is also known as default risk. It is related to a loss that may occur from the failure of repayment of a credit advance made by a bank to its customers. Credit risk is inherent to the business of lending funds. It may lead to losses when banks' customers experience deterioration in financial condition, making it impossible to recover principal and interest on loans, securities and other monetary claims outstanding. Management of this type of risk is the most fundamental task in banking operations. The objective of credit risk management therefore is to minimize the risk and maximize bank's risk adjusted rates of return by assuming and maintaining credit exposure within the acceptable parameters.
- Liquidity risk which refers to the ability of a bank to fund increases in assets and meet obligations as they come due, without incurring unacceptable losses. Liquidity risk therefore occurs when a bank is not able to meet the payment of commitment it has made. To manage liquidity risk, the banks must periodically examine the structure of fund sources and uses and implement measures needed to improve this structure.
- Market risk arises when financial institutions actively trade assets and liabilities (and derivatives) rather than holding them for longer-term investment, funding or hedging purposes. Market risk implies not merely the risk of loss but also the potential for gain from interest rate, equity return and foreign exchange risk in that as these risks increase or decrease, the overall risk of the financial institution is affected. Market risk could therefore be defined as the incremental risk incurred by a financial institution when interest rate, foreign exchange and equity return risks are combined with an active trading strategy, especially one that involves short trading horizons such as a day. Market risk can be in the form of interest rate risk, refinancing risk and reinvestment risk. It is therefore required that bank should conduct strict management and control of market risk based on the awareness that the possibility of substantial losses is inherent in the nature of market transactions.
- The Bank of International Settlement (BIS) (1997), views operational risk (inclusive of technology risk) as the risk of loss resulting from inadequate or failed internal processes, people and systems or from external events. The breakdown of internal controls and corporate governance leading to "error, fraud, performance failure, compromise on the interest of the bank resulting in financial loss". The

objective of operational risk management is to find out the extent of financial institutions operational risk exposure; understand what drives it; to allocate capital against it; and identify trends internally and externally that would help in predicting it.

- Strategic risk is viewed by Onyekwelu and Onyeka (2014) as the risk arising from the overall strategy of the bank that includes the quality of the strategic planning process and the implications of such strategy especially for risk appetite and track record of implementation. It explains risk associated with business targets as risk from bank's product, services and customers.

It is clear from the risks that are said to be managed by banks above that technology related risks have not received prominence. This is probably because of the manner in which we have traditionally viewed banks. Of importance here is that whatever package or transaction that banks offer today, technology has an important role. In developing countries such as Nigeria where risks have materialized and caused vulnerabilities in the banking system, we are of the view that exploiting technology in order to reduce missed opportunities and realized risks, and using this technology in order to integrate different systems and collect and analyze massive volumes of data from an unlimited number of sources across multiple locations for the benefit of stakeholders is important.

In highlighting the importance of adequate management of technological risk, the Centre of Excellence in Financial Services (COEFS, 2018) stated how in this era of the fourth industrial revolution, technological advancements like artificial intelligence means that human and digital systems can interact more profoundly than ever before. Applying this technology in financial services has the potential to reduce costs and improve efficiency, allow customers to transact seamlessly and in real time, and improve providers' understanding of customer behavior and needs, allowing for the personalization of financial services. On the other hand, the fast pace of technological innovation and adoption within financial services means that banks will increasingly face a common set of operational risks around new technology. If these risks are not properly managed by banks, they have the potential to create instability and undermine consumer trust in the banking system as instances of cybercrime, systems failure and compromised data increase in frequency.

Particularly, as the guardians of sensitive private and financial data, fraud and cybercrime is the greatest reputational risk facing banks in the digital age. One obvious reason contributing to the intensity of this risk is that banks have a number of fronts they need to protect. The digitization of bank channels and underlying processes create multiple avenues through which virtual attacks can take place, notwithstanding the need to protect banks' physical premises and systems from intrusion. In addition to banks' internal systems and processes, the increasing use of outsourced service providers and the integration of banks' systems with external vendors adds another layer of vulnerability. It is difficult for banks to cover all of these fronts comprehensively (COEFS, 2018). All of these require banks to develop comprehensive cyber and data resilience strategies. Failure in this regards in the information revolution era could launch the banking sector into another round of crisis that may be difficult to recover from.

Previously, the digitization efforts of banks have been customer-centered (such as online marketing) and the operations that support those efforts (customer onboarding, customer servicing). Only recently have banks expanded their transformations into other parts of the organization, including the risk function. In this regard, it is clear that banks concede that there is value in deploying technology as part of managing risks.

A survey conducted by the Institute of International Finance/McKinsey shows that 70 percent of respondents reported that senior managers are paying moderate attention to risk-digitization efforts; 10 percent say that senior managers have made these efforts a top priority. Risk digitization is clearly an established topic in the executive suite (Institute of International Finance/McKinsey, 2018). The survey lays out seven key elements a bank must develop for a digital risk transformation, namely:

- Accordingly, enhanced data governance and operating models are likely to improve the quality of the data, make risk and business decisions more consistent, and ensure responsiveness to risk's data needs. The report is of the view that one important enhancement is the need to consider data risk as a key element of the risk taxonomy, linked to a specific risk-appetite statement and data-control framework and to accommodate far more varieties of data. The survey found that approximately 30 percent of the respondents say that new data sources will probably have a high impact on their work. This is in line with Agrawal et al.'s (2018) assertion that the power of artificial intelligence will allow resources to be focused somewhere else whereas technology will focus on predictions.

- On process and workflow automation, it is expected that as risk automates tasks such as collateral data entry, often through robotic process automation (RPA), it can combine several of them into smart workflows: an integrated sequence performed by groups of humans and machines across an entire journey (e.g., credit extension fulfillment). In addition to greater efficiency, smart workflows create a more seamless and timely experience for customers. The survey found that about a quarter of respondents believe that more than 15 percent of costs can be cut across different risk disciplines, except in credit, where the number is a bit above 60 percent. Around 30–45 percent of respondents see 5–15 percent cost-reduction potential from automation, depending on risk type. Ninety percent see benefits from increased precision and 55 percent believe automation will improve compliance with regulation. As a knock-on effect, risk resources will focus more on the value-adding activities they have been trained for. And 84 percent of respondents expect an increase in customer and employee satisfaction.
- On advanced analytics and decision automation, the survey is of the view that sophisticated risk models (e.g., those built on machine-learning algorithms) can find complex patterns (such as sets of transactions indicative of invoice fraud) and make more accurate predictions of default and other risk events. The survey found that nearly three-quarters of risk managers surveyed expect advanced analytics to have a significant impact on their work, while 50 percent say credit decision times will fall by 25–50 percent and a few respondents even believe that times could fall by 75–100 percent. This is in line with Agrawal et al.'s (2018) assertion that the power of artificial intelligence is on prediction, that is, it will make prediction accessible and faster.
- With regard to a cohesive, timely and flexible infrastructure, if risk infrastructure is digitized, it will evolve to support several other building blocks: innovative data-storage solutions, new interfaces, easier access to the vendor ecosystem, and so on. It will use techniques like application as a service, obtained from application service providers (even on open banking platforms). Approximately 45 percent of the respondents see innovative technologies as a high-impact building block. "No code" and "low code" solutions will put control further in the hands of risk executives and reduce the number of end-user computing tools. Nearly 60 percent of the respondents expect innovative data-storage structures to have a significant impact on risk management.

- On smart visualization and interfaces, risk digitization will deliver its insights in more intuitive, interactive and personalized ways through risk dashboards, augmented-reality platforms for customers, and other interfaces. The survey found that nearly 20 percent of risk managers expect nascent technologies, such as augmented reality, to have a high impact.
- On external ecosystem, risk digitization will facilitate easy partnerships with external providers to vastly improve customer on-boarding, credit underwriting, fraud detection, regulatory reporting and many other activities. The survey found that two-thirds of respondents see fintechs more as enablers than disruptors, while 63 percent of North American respondents plan to use industry utilities to deal with regulatory burdens.
- On talent and culture, risk digitization is expected to have a far greater share of digital-savvy personnel with fluency in the language of both risk and the business, operating within an agile culture that values innovation and experimentation. The survey reports that new profiles seen as most critical in a digitized risk function include data scientists and modeling experts. Accordingly, many risk leaders think that their teams will need to develop these skills rather than hire non-risk professionals and expect them to learn risk.

Empirical Analysis of the Current State of Risk Management of Banks in Nigeria

Given the recent challenges faced by the banking sector in Nigeria, we are of the view that the evolutions in computing and risk technology which has led to the developments in technologies could be used to exploit big data, conduct complex analytics and these technical advancements could offer the Nigerian banks and authorities such as the CBN better abilities for enhancing risk management effectiveness thereby reducing vulnerabilities and exploiting opportunities that could arise.

We set to investigate whether or not the Nigerian banking sector was exploiting technology in order to reduce missed opportunities and realized risks, whether technology has been exploited in order to integrate different systems and collect and analyze massive volumes of data from an unlimited number of sources across multiple locations. The results of content-analyzing integrated reports of selected Nigerian banks where the focus had been on the manner and the way in which risks have been captured and the determination whether Nigerian banks indicate in these reports whether they have deployed technology or not and whether or not the advanced technology would have been deployed in order to reduce missed opportunities and realized risks.

Table 14.1 presents the empirical investigation of 13 banks that rank within the top 40 firms in Nigeria in terms of market capitalization. The latest annual reports of the sampled banks were content-analyzed to extract information on the current risk management approach. Basically, four aspects of risk management approach were examined, namely, the specific risk management framework guiding the risk activities of the bank; the presence of risk committee whose responsibility is to provide oversight functions on behalf of the Board; basic technology employed to manage all forms of risks in the banks; and the application of advanced ICT for risk management in the information revolution era.

The first three aspects are fully manifest in the sampled banks. These are regarded as the traditional approach to managing risks in the banking industry and have existed for some time. Only two banks appear to be proactive in their risk management approach. Although the depth of risk digitization in these two banks are not verified, the latest annual reports issued by them have some elements of these advanced ICT technologies for managing risks.

Table 14.1 Current risk management approach in Nigerian banks

SN	*Banks*	*Risk management framework*	*Risk committee*	*Basic ICT for risk management*	*Advanced ICT for risk management*
1	GTBank	√	√	√	
2	Zenith	√	√	√	
3	Stanbic	√	√	√	
4	FBN	√	√	√	√
5	UBA	√	√	√	
6	ECOBANK	√	√	√	
7	Access	√	√	√	√
8	Union	√	√	√	
9	Fidelity	√	√	√	
10	Sterling	√	√	√	
11	FCMB	√	√	√	
12	Diamond	√	√	√	
13	Wema	√	√	√	
	Total	*13*	*13*	*13*	*2*

Conclusion, Recommendations and Policy Implications

It is clear in the integrated reports that banks have not embraced advanced technology for the purpose of managing risk. To reduce the likelihood of the reoccurrences of recent challenges faced by the banking sector in Nigeria, we are of the view that the evolutions in computing and risk technology which has led to the developments in technologies could be used to exploit big data, conduct complex analytics and these technical advancements could offer the Nigerian banks and authorities such as the CBN better abilities for enhancing risk management effectiveness thereby reducing vulnerabilities and exploiting opportunities that could arise.

With the rise of technology, majority of banking activities are already executed through technological instruments. As a policy recommendation, in addition to the traditional risk and audit committees that are clearly outlined in the code, the CBN would have to incorporate ICT committee in the codes of corporate governance for banks in Nigeria. Alternatively, the code should make it mandatory that some of the external members of the audit and risk committee should possess technological competencies.

References

Adeoye, A., & Amupitan, M. D. (2014). Corporate governance in the Nigerian banking sector: Issues and challenges. *European Journal of Accounting Auditing and Finance Research, 3*(5), 64–89.

Agrawal, A., Gans, J., & Goldfarb, A. (2018). *Prediction machines: The simple economics of artificial intelligence*. Boston, MA: Harvard Business Review Press.

Akande, O.B. (2016). *Corporate governance issues in the Nigerian banking industry*. PhD dissertation. Walden University.

Akinkoye, E. Y., & Olasanmi, O. O. (2014). Corporate governance practice and level of compliance among firms in Nigeria: Industry analysis. *Journal of Business and Retail Management Research (JBRMR), 9*(1), 13–25.

Alushna, M. (2017). *Corporate governance best practice: Tasks and shortcomings. Responsible corporate governance towards sustainable and effective governance structure*. Springer.

Aransiola, S. Y. (2013). The impact of consolidation on profitability of commercial banks in Nigeria. *Financial and Quantitative Analysis, 1*, 15–25. https://doi.org/10.12966/fqa.05.01.2013

Ayodele, T. O., & Alabi, R. O. (2014). Risk management in Nigeria banking industry. *Research Journal of Finance and Accounting, 5*(2), 131–136.

Cadbury Report. (1992). *Report of the committee on the financial aspects of corporate governance*. Retrieved September 12, 2018, from http://www.ecgi.org/codes/documents/cadbury.pdf

Central Bank of Nigeria (CBN). (2014). *Code of corporate governance for banks and discount houses*. Abuja, Nigeria.

CBN. (2019). List of financial institutions. Retrieved from https://www.cbn.gov.ng/supervision/Inst-DM.asp.

COEFS. (2018). *The impact of the 4th industrial revolution on the South African financial services market*. Retrieved October 12, 2018, from https://www.genesis-analytics.com/uploads/downloads/COEFStheimpactofthefourthindustrialrevolutiononfinancialservicesinSouthAfrica-final-1-FR.pdf

Ernst & Young. (2014). Retrieved from https://www.ey.com/Publication/vwLUAssets/EY-leveragingtechnology-and-data-for-cost-effective-risk-management/$File/EYLeveraging-technology-and-data-for-cost-effective-risk-management.pdf.

Harari, Y. N. (2018). *21 lessons for the 21st century*. London: Jonathan Cape.

Hempel, G. H., Coleman, A. B., & Simonson, D. G. (1990). *Bank management: Text and cases* (3rd ed.). New York: John Wiley & Sons, Inc.

IIF/McKinsey. (2018). The future of risk management in the digital era. Digital risk survey of 35 banks, companies, and other institutions. Retrieved March 30, 2018, from https://www.mckinsey.com/~/media/McKinsey/Business%20Functions/Risk/Our%20Insights/The%20future%20of%20risk%20management%20in%20the%20digital%20era/Future-of-risk-management-in-the-digital-era-IIF-and-McKinsey.ashx

Knopjes, B. (2017). Why IoT is the future of integrated risk management. Retrieved October 12, 2018, from https://www.isometrix.com/iot-in-risk-management/

Larcker, D., & Tayan, B. (2011). *Corporate governance matters: A closer look at organizational choices and their consequences*. Upper Saddle River: Pearson Education.

Liu, V. (2004). Environmental considerations and business operations of commercial banks in China—A case study of the project loan appraisal policy of ICBC. Lund, Sweden: IIIEE.

Mallin, C. (2004). *Corporate governance*. Oxford: Oxford University Press.

Marwala, T., & Hurwitz, E. (2017). *Artificial intelligence and economic theory: Skynet in the market*. Springer.

Moloi, T. (2009). *Assessment of corporate governance reporting in the annual report of South African listed companies*. Masters dissertation, University of South Africa.

Moloi, T. (2014). Leading external and internal indicators of credit risk in the top South African banks. *Journal of Risk Governance and Control: Financial Markets and Institutions, 4*(3), 51–65.

Moloi, T. (2016a). The nature of credit risk information disclosed in the risk and capital reports of the top-5 South African Banks. *Banks and Bank Systems, 11*(3), 84–90.

Moloi, T. (2016b). A cross sectoral comparison of risk management practices in South African organizations. *Problems and Perspectives in Management, 14*(3), 99–106.

Moloi, T. (2018). Analysing the human capital capabilities in the enterprise risk management function of South Africa's public institutions. *Business and Economic Horizon, 14*(2), 375–388.

Monnin, P. & Jokipii, T. (2010). *The impact of banking sector stability on the real economy.* Swiss National Bank. Working group paper.

Nzotta, M. S. (2002). *Corporate financial decisions.* Owerri: Oliverson Industrial Publishers.

OECD. (2004). *Principles of corporate governance.* Retrieved October 9, 2018, from http://www.Oecd.Org/Corporate/Ca/Corporategovernanceprinciples/31557724.Pdf

OECD. (2015). G20/OECD *principles of corporate governance.* Retrieved October 9, 2018, from http://www.oecd.org/daf/ca/Corporate-Governance-Principles-ENG.pdf

Onyekwelu, U. L., & Onyeka, V. N. (2014). Financial risk management: A review of the role of the central bank of Nigeria. *Research Journal of Finance and Accounting, 5*(5), 40–52.

Owualla, S. I. (2000). *Principles of financial management.* Lagos: G. Mag Investment Ltd.

Patterson, T. (2015). *The use of information technology in risk management.* IBM.

Sanusi, L.S. (2011). The impact of the global financial crisis on the Nigerian capital market and the reforms. *7th Annual Pearl Awards and Public Lectures.* Muson Centre, Onikan. Retrieved from http://www.cbn.gov.ng

Tennant, D. F., & Tracey, M. R. (2014). Financial intermediation and stock market volatility in a small bank-dominated economy. *Journal of Developing Areas, 48*, 73–95. https://doi.org/10.1353/jda.2014.0073

The Bank of International Settlement (BIS). (1997). Principles for sound management of operational risk. Retrieved from https://www.bis.org/publ/bcbs195.pdf.

The Punch (2017, September 17). *Tackling weak corporate governance in banks.* Retrieved October 9, 2018, from https://punchng.com/tackling-weak-corporate-governance-in-banks/

Tricker, B. (2012). *Corporate governance: Principles, policies and practices.* Oxford: Oxford University Press.

Uwuigbe, O. R. (2011). *Corporate governance and financial performance of banks: A study of listed banks in Nigeria.* Ota: Covenant University.

CHAPTER 15

Digital Technology to Enhance Project Leadership Practice: The Case of Civil Construction

John Ekechukwu and Thorsten Lammers

Introduction

In business, the last few decades have belonged to a certain type of character or worker. A manager who initiates, plans, executes, monitors and controls. These people may be seen as reactive “firefighters”, who put out flames, focus on completing tasks and respond to unfolding events (Barber & Warn, 2005). However, the future belongs to a very different kind of person, with a different kind of mind. Leaders, who are “creators, empathizers, pattern recognizers, and meaning makers” (Pink, 2006). These people are proactive “firelighters”, who create change, envision the future and evoke passion (Barber & Warn, 2005). And as the complexities of organizational projects become more volatile, the demands for these characters heighten. As such management now also implies leadership.

J. Ekechukwu (✉)
Turnbull Engineering, Sydney, NSW, Australia
e-mail: John.Ekechukwu@alumni.uts.edu.au

T. Lammers
University of Technology Sydney, Sydney, NSW, Australia
e-mail: thorsten.lammers@uts.edu.au

B. George, J. Paul (eds.), *Digital Transformation in Business and Society*, https://doi.org/10.1007/978-3-030-08277-2_15

Leadership, however, is conventionally associated with organizations, not projects (Cavaleri & Reed, 2008). Its theories and respective characteristics are rooted from the context of permanent organizations (Tyssen, Wald, & Spieth, 2014). These lasting forms of organizing facilitate a sense of autonomy and unbounded freedom that may enable leaders to imagine alternative visions for the future (Cavaleri & Reed, 2008). Yet, projects do not have the same magnitude of autonomy. A project environment of limited and predefined duration creates a sense of short-term orientation and hampers the development of deeper social relations such as trust and camaraderie (Tyssen et al., 2014). Additionally, a project evokes focus toward the management of daily objectives and immediate achievements, thus inhibiting actions of leadership and long-term vision (Toor & Ofori, 2008). Each project is unique and conditions for team selection and motivation are often far from ideal (Anantatmula, 2010). At times, project members are engaged in more than one project, causing additional burdens on focus and commitment (Anantatmula, 2010). Because of this, role ambiguity becomes an issue as well as the uncertainty of allocated control (Zhang & Fan, 2013).

Therefore, the role of a "project leader" is significantly different to that of an "organizational leader". Leading an organization allows for freedom, whereas leading a project, involves fulfilling a "promise" to a customer, ultimately constricting a leader to operate within the confines of such promise (Cavaleri & Reed, 2008). In this sense, a project leader is one who practices leadership within predetermined parameters. This makes the role of a project leader a rare commodity that is valued now more than ever before. So far, literature has focused on endorsing particular leadership behaviors and styles, while ignoring the difficulties faced by professionals in practicing these behaviors.

This study will explore digital transformation as a mechanism to facilitate leadership development using the case of the Australian civil construction industry. The focus is to shift the emphasis of digital advancements toward the social phenomenon of leadership. Digital technology alters the way people communicate, interact and relate to one another (Nagy & Koles, 2014). To neglect its potential to enhance project leadership development may be a substantial oversight in the evolution of leadership practice. To date, there have been few attempts to explore digital technology as project leadership technology, particularly within an industry such as construction, in dire need of support to elevate project managers into project leaders.

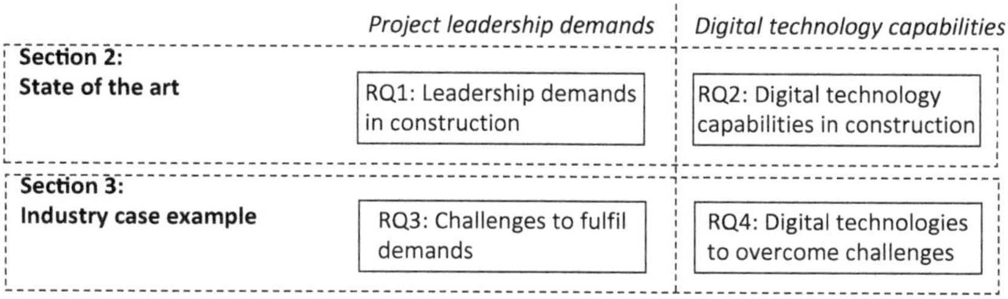

Fig. 15.1 Research design overview

This chapter aims to explore the potential for digital technology to enhance project leadership practice in construction projects. To investigate this potential, we will shed light on the following questions:

1. What are current leadership demands in construction?
2. What are current digital technological capabilities relevant to construction?
3. What challenges are construction professionals facing in meeting the leadership demands?
4. What is the feasibility of digital technology assisting professionals in overcoming these challenges?

The methodology employed to answer these questions involves a literature review coupled with a case example from the Australian civil construction industry (see Fig. 15.1).

Leadership Demands and Technological Capabilities in Construction

In the following section, we provide an academic assessment of the existing body of knowledge regarding leadership demands in the construction industry as well as capabilities of new digital technologies relevant for the construction sector. The focus is to discover important insights that can be used to guide and facilitate the investigation into the industry case example. In reviewing existing literature on those topics, the criteria listed in Table 15.1 have been established to ensure that relevancy and validity is maintained throughout the study.

Table 15.1 Overview of literature review

Source	The source is attainable in digital form.
Content	Content principally focuses on leadership and/or technology.
Keywords	Search is conducted using a combination of relevant words that depict leadership/technology (including plural or connotative equivalents).
Context	Sources are published within an international journal in the domain of construction, engineering, project management and the built environment.
Reliability	The source is academic/scholarly.
Relevancy	The source is published in English language within the date range specified (RQ = Research Question): • RQ1: sources published within the last ten years (from 2007 onward) • RQ2: sources published within the last three years (from 2014 onward)

Leadership Demands in Construction

Leadership challenges are amplified within the construction sector, an industry that will spend US$97.7 trillion globally in the next decade and account for 13.2% of world GDP by 2020 (Richardson, 2017). Construction projects are often large, multi-disciplinary and time-limited undertakings that exacerbate coordination and communication, as well as impose a great degree of uniqueness in setup, activities and content (Tyssen et al., 2014). Yet without tangible support, education and/or training, project managers are expected to "select, equip, train and influence" followers with diverse gifts, abilities and skills while operating within predefined parameters of time, budget and scope (Ameh & Odusami, 2014). The industry is suffering from a leadership crisis. Not from the professionals themselves, but from the disparity between the demands in project leadership and the investment and research in leadership development and training (Toor & Ofori, 2008). There is too much focus on the outcomes of leadership in neglect of the means in delivering these outcomes. Therefore, project managers need assisting tools and mechanisms that can balance the playing field and enable them to maximize their project leadership potential. As Khan (2015) claims, 80% of project failures are a result of poor project leadership or lack of project leadership skills. Management and technical competency alone are no longer sufficient for project success (Schuhmann, 2010). The following results highlight the variety of leadership demands relevant to the construction industry.

A US case study conducted by Cavaleri and Reed (2008) utilizes Seivert's five elements model to present a leadership strategy for improving project performance. The model constitutes of "Essence" (defining purpose and values), "Air" (planning, analyzing and envisioning new possibilities), "Fire" (motivation and passion), "Water" (team building, loyalty, collegiality and camaraderie) and "Earth" (structure and policies). The expectation is that the project leader must be sensitive to the needs facing the team at the time and find a balance between the five elements. In teams in which the leader places emphasis on "Air" and "Fire" in neglect of "Water" and "Earth", projects are seen to fall behind, because important details that are necessary for execution are ignored. In a similar vein, when a project leader focuses on "Fire", "Water" and "Earth", while overlooking "Essence" and "Air", teams are likely to veer off in a direction that is misaligned with the project mission and objective.

Similarly, a study conducted by Patterson (2010) uses a case study of the Oresund Bridge construction (linking the cities of Copenhagen, Denmark and Sweden) to present a five-E model of leadership. "Envision", "Engage", "Energize", "Enable" and "Execute". By envisioning, "leaders see what the future could be rather than what it is". By engaging, leaders build relationships and seek collaboration. By energizing, leaders inspire and motivate. By enabling, leaders provide the necessary resources to facilitate followers in completing objectives. And by executing, leaders combine all elements together successfully.

These elements are further emulated in a study conducted by Slattery and Sumner (2011), who utilize the Leadership Practice Inventory (LPI) developed by Kouzes and Posner (2003) to examine the most favorable leadership characteristics among a group of 151 high potential construction project managers. The LPI purports to measure the ability of an individual to model the way (clarify values and set the example), inspire a shared vision (envision possibilities), challenge the process (take initiative, innovate and improve), enable others to act (foster collaboration, trust and facilitate relationships) and encourage the heart (provide emotional support).

Furthermore, surveys and structured personal interviews of 69 engineering professionals conducted by Anantatmula (2010) were used to develop a project management leadership model that consists of significant people-related leadership factors that influence project performance. These factors include: Create clarity in communication, define roles and responsibilities, communicate expectations, employ consistent processes, establish trust, facilitate support and manage outcomes.

The demand for authenticity and trust is accentuated throughout literature. Toor and Ofori (2008) conduct a literature review based in Singapore to rationalize the need for Authentic Project Leadership Development (APLD) for construction project leaders. The study explores four components of authenticity; "Awareness" (trust in one's motives and cognitions), "Unbiased Processing" (refraining from denying or distorting private knowledge), "Behavior" (acting in accord with one's values) and "Relational Orientation" (openness and truthfulness in relationships). Findings from this study argue that authentic leaders must know themselves, define their values, understand their motivations, build their support team, and "stay grounded by integrating all aspects of their life". Authenticity demands positive energy, integrity, moral character, self-discipline, optimism, resilience, clear purpose, concern for others and personal values.

As articulated by Kappagomtula (2017), without trust, leadership cannot function. Using investigative surveys of 36 experts in leadership across China and India, Kappagomtula models an analogy of a leadership "wheel" which must be maintained within cross-cultural and multicultural project teams. The model stipulates that the hub of the steering wheel is "Trust". The spokes of the wheel are "Loyalty", "Communication", "Integrity", "Competency" and "Consistency". The wheel itself represents "Culture", and the lubricant for the wheel is "Conflict Management". The emphasis of this analogy is the need for the leader to understand its components (the hub and spokes), maintain its integrity (lubricant) and steer the project team toward the destination of the drive. However, the leader must also learn when to let go of the "wheel" and surrender control.

This notion is also explored by Clarke (2012) who develops a theoretical understanding of shared leadership within projects. He argues that no individual performs all leadership functions, but rather members of the team collectively contribute to the leadership system. Therefore, a leader must focus on enriching established connections between team members and developing new connections to create a cohesive system in which leadership is distributed across all members.

Multiple studies have identified the need for project leaders to embody different leadership styles based on the varying stages of construction projects. Arain (2012) uses pilot questionnaires and interviews of 78 project management professionals working on the Suvarnabhumi

Airport (the largest construction project in Thai construction history) to develop a Quadrilateral Model of project leadership. The model focuses on a leader's ability to balance between four key leadership styles: "Coaching" (aligning goals and facilitating interaction among a team), "Connecting" (listening, empathizing, engaging and revering), "Controlling" (implementing rules) and "Charismatic" (supporting, inspiring and selling a vision). The research posits that leaders need variations in their leadership styles, at different stages of the project, until a balance between personal values, concerns for tasks and concern for people is maintained.

This concept is also explored in interviews across three Sri-Lankan construction projects conducted by Senaratne and Samaraweera (2015). This empirical study highlights the four stages of team development within projects: "Forming", "Storming", "Norming" and "Performing". The authors posit that a construction project manager should employ different leadership roles based on the needs and behaviors of the team members of each stage. For example, a leader should break down interpersonal barriers and establish team goals in the forming stages, but should only intervene where necessary in the performing stage.

This demand of leadership style adaptability is further reinforced by Oke's (2010) Multifactor Leadership Questionnaire (MLQ) of 57 Nigerian construction professionals. Findings from this study demonstrate that construction project leaders should possess the ability to adapt different leadership styles depending on the situation at hand. Leaders should embody a relations-oriented style at the beginning of a project, while adopting a much more task-oriented style during the closing of the project.

Leadership demands have also focused on the need for positive psychology and emotional intelligence among construction professionals. Emil Berg and Terje Karlsen (2014) conduct in-depth interviews with three experienced project managers to accentuate the need to encourage and develop positive emotions among followers. The study reveals how "Positive Meaning" (creating something), "Positive Emotions" (humor and optimism), "Positive Relations" (collaboration) and "Signature Strengths" (emotional intelligence) can create positive results within a project. These emotional elements play a major role in the development and trajectory of relationships within project settings (Pryke, 2015).

In a study, conducted by Pryke (2015), which uses surveys and live observations of 68 construction professionals, leaders are described as "managers of group emotions". Findings in this study demonstrate a clear correlation between emotional intelligence and project performance and further suggest that "Emotional Sensitivity" (ability to sense and interpret the non-verbal messages of others) and "Emotional Expressiveness" (ability to express emotions) are fundamental ingredients for "leader-follower chemistry". The expectation is that a leader should create a harmonious, smooth and in-tune connection between members of the team through emotional exchange and interaction.

Zhang and Fan (2013) further lament on emotional intelligence as a leadership requirement in their questionnaire-based survey covering 112 Chinese project managers. They find that characteristics of "Self-awareness", "Self-confidence", "Self-control", "Empathy", "Positivity" and "Cultural understanding" are key ingredients for project performance.

This is again emphasized by Galvin, Gibbs, Sullivan, and Williams (2014), who survey 38 project managers to define leadership competencies of "Emotional Resilience", "Intuitiveness", "Interpersonal Sensitivity", "Motivation" and "Conscientiousness". The study further postulates that an effective project leader must possess varying levels of sensitivity to political, technical and emotional elements of a project.

This is supported by Müller and Turner (2010), who use a web-based "Leadership Dimensions Questionnaire" on 400 project managers around the world to emphasize the need for project leaders to not only develop "Emotional Intelligence", but to balance between "Emotional", "Managerial" and "Intellectual Competencies".

Researchers have even attempted to appropriate religious teachings into necessary spiritual demands of leadership. Yngvason (2013) reviews the four gospels of the Bible to propose how project leaders can enhance their own practices by embracing the teachings of Jesus. The study proposes that Jesus utilized leadership tactics that enabled him to create a sense of purpose for a small number of followers that eventuated into a way of life for billions. The authors assume that like a project, the initial goal of spreading "the good news" was a planned endeavor, in which Jesus demonstrated transformational characteristics that included; "Individualized Consideration" (attentiveness to followers needs), "Intellectual Stimulation", "Inspirational Motivation" and "Idealized Influence" (a role model for ethical behavior).

Similarly, in a case study of eight Hungarian construction companies conducted by Lazányi and Dóka (2015), it is suggested that teachings from the Old Testament provide clues on leadership behavior. Leaders should be worthy, righteous men, who refrain from covetousness.

This is further explored from an Islamic perspective by Senam, Rashid, Sarkawi, and Zaini (2014), who review the Quran to highlight the need for authentic, servant and ethical leadership practices among construction professionals. The study unearths spiritual and moral values of leadership and emphasizes the need for leaders to carry out leadership activities with minimal waste or destruction and the upmost of ethical integrity.

Table 15.2 summarizes the demands identified in the above empirical studies of leadership relevant to the construction industry and groups them into three focus groups.

Digital Technology Capabilities Relevant to Construction

Digital transformation is the "application of digital technologies to fundamentally impact all aspects of business and society" (Gruman, 2016). The digital component involves technology, while the transformation component involves people (Del Rowe, 2017). It can enable individuals and organizations to transcend from one way of working to another.

This section focuses on reviewing the technological opportunities relevant to the construction industry. It provides a non-exhaustive list of technology examples, current and future trends of digital technology and their capabilities and application potentials for the construction industry.

Smartphones

Smartphones serve as the hubs of interconnectivity (Ahmed & Ahmad, 2014). They allow for knowledge sharing and exchange, as well as collaboration through a plethora of functionalities and applications. Smartphones can be used as a single mode of analysis for digital technology such as augmented reality (AR), to aid learning, understanding and remote collaboration (Joan, 2015); computer-aided design (CAD), to clarify important design details (Moreno, 2014); near-field communication (NFC), to track and monitor construction worker health and safety

Table 15.2 Overview of key leadership demands in construction

Leadership demand	*Source*						
	Cavaleri and Reed (2008)	*Patterson (2010)*	*Slattery and Sumner (2011)*	*Anantatmula (2010)*	*Toor and Ofori (2008)*	*Kappagomtula (2017)*	*Clarke (2012)*
Team focus							
Empower relationships	•	•	•		•		•
Enable effect. communicat.				•		•	
Establish trust & loyality			•	•	•	•	
Provide emotional support			•				
Align team's goals							•
Leader focus							
Envision new possibilities	•	•	•				
Give control to team							•
Lead by example	•	•	•			•	
Manage conflicts						•	
Balance all demands	•	•				•	
Project focus							
Define work struct. & policy				•			
Define purpose and values	•		•		•		
Communicate expectations				•			
Provide necessary resources		•		•			
Adapt style to project phase							

Arain (2012)	*Senaratne and Samaraweera (2015)*	*Oke (2010)*	*Emil Berg and Terje Karlsen (2014)*	*Pryke (2015)*	*Zhang and Fan (2013)*	*Galvin et al. (2014)*	*Müller and Turner (2010)*	*Yngvason (2013)*	*Lazányi and Dóka (2015)*	*Senam et al. (2014)*
	•	•	•	•						
•										
					•					
•			•	•	•		•	•		
•	•									
•			•					•		
	•									
•			•	•	•			•	•	•
•							•			
•		•								
•										
•		•								

(Akhavian & Behzadan, 2016); and credential verification services (CVS), to monitor employee training records and renewals. Smartphones are an extension of their users, and with the many applications that can be integrated within these devices, they act as a remote for digital transformation (Azhar, Jackson, & Sattineni, 2015).

Wearable Technology

Wearable technology (or "wearables") describes the integration of smart electronics within items that may be worn or implanted (Wright & Keith, 2014). Smart glasses, for example, incorporate an optical display that can advance information exchange at construction sites (Moon & Seo, 2015). In a split second, electronic documents, video and images can be shared, saved and retrieved (Heembrock, 2015), allowing construction professionals to be freed from the burden of holding paper documents, such as drawings and specifications (Moon & Seo, 2015). Wearable wristbands that measure sleep patterns and circadian rhythm can aid in managing unsafe working conditions by identifying fatigue, illness and other physical risk factors (Cullen, 2016). These functions can also be extended to mental health and well-being, by reminding construction professionals to stretch, hydrate and replenish (Heembrock, 2015). With the integration of woven sensors, wearable clothing can inform workers if they are exceeding lifting thresholds, have been overexposed to UV rays or are at risk of a slip and fall incident (Heembrock, 2015). Self-charging shoes that detect temperature and track location and motion can also ensure worker well-being is maintained when working alone (Linder, 2017). With the variety of sensors that can be integrated within clothing or accessories, leaders can understand the well-being of their team and advance their role in improving health and safety within construction projects (Linder, 2017).

Drones

Drones function as aerial imaging vehicles that can transcend data collection capabilities within construction projects. They allow for site progression and operations to be accurately tracked and reported (Dillow, 2016) and can enhance efficiency with site surveys, photography and mapping. With sensory features integrated, drones can turn data into 3D structural models, topographical maps and volumetric measurements (Dillow, 2016), which can translate to enhanced productivity, safety, quality and security within projects. Additionally, "smart drones" possess the capability of recognizing objects and can incorporate a "sense-and-avoid-system" to self-monitor project operations (Hambling, 2015). These aerial capabilities add a greater dimension to project tracking and can allow project leaders to effectively communicate, mitigate potential issues, reduce costs and limit delays (Dillow, 2016).

Virtual Reality, Augmented Reality and Mixed Reality

Advanced visualization technologies can impact upon the way groups and individuals collaborate, communicate and engage (Phelps, 2014). Augmented reality (AR) overlays the visible natural world with a layer of digital content, while mixed reality (MR) integrates virtual objects into the natural world (Heiskanen, 2016). These capabilities for digital content to interact with the real world enable new possibilities for safety, planning, quality and management (Heiskanen, 2016). AR used through smart glasses can enable for a digitally enhanced view of the world that can facilitate the learning and awareness for student and professional development in construction (Kivrak, Arslan, & Tuncan, 2015; Ahmed & Ahmad, 2014). For example, the Smart Helmet by American augmented reality company DAQRI creates an intuitive experience as it connects users to the work environment and provides relevant information instantaneously (Heiskanen, 2016). Additionally, holographic technology such as Microsoft Hololens can allow remote teams to experience the spatial presence of the design model, collaborate and share information instantaneously over three-dimensional holograms (Heiskanen, 2016).

Virtual reality (VR) however creates a completely simulated environment (Sampaio & Viana, 2014) and can allow for four-dimensional viewing and interaction with all stages of the construction process. Project team members can therefore optimize planning and design and create an interactive environment for construction resourcing and scheduling VR can also be used to support learning and awareness for safety and risk management and develop a platform for users with limited knowledge to easily simulate the activities needed (Kassem, Benomran, & Teizer, 2017). These visual capabilities can aid in prediction and decision making by allowing users to view, interact and simulate their projects throughout its life cycle (Sampaio & Viana, 2014).

Building Information Modeling (BIM)

BIM is a three-dimensional (3D) model-based technology process for creating and managing project information (Succar, 2009). BIM will reduce the "blind spots" of two-dimensional (2D) construction drawings (Lee et al., 2015) and facilitate for extensive stakeholder collaboration (Kim et al., 2016). It can improve resource coordination, quality control, information exchange, safety, security, decision making and reduce manual tasks (Park, 2016). However, its collaborative platform (Lee et al., 2014) is what can facilitate project leaders to effectively communicate in real-time among project stakeholders (Park, 2016). Furthermore, new advancements of four-dimensional (4D) BIM and five-dimensional (5D) BIM enable further improvements in areas such as design, information management and sustainability. 4D BIM incorporates time to establish links between project activities and scheduling, which enables construction professionals to conduct constructability checks before construction (Kim, 2016). Whereas 5D BIM incorporates costing within the model to ensure the effective management of costs throughout a project life cycle, including cost estimation and budgeting (Kim, 2016).

Artificial Intelligence and Software

Artificial intelligence (AI) can redefine how processes are undertaken within the construction industry (Blanco, Fuchs, Parsons, & Ribeirinho, 2018). Collisions and inefficiencies within models such as BIM can be preempted, tracked and controlled (Garcia De Soto & Adey, 2016). Knowledge contained in daily reports, schedules, weather forecasts, and many more can be transformed into a living knowledge base and means of deductive and plausible conclusions (Klashanov, 2016). Garcia De Soto and Adey (2016) also believe that by integrating the case-based reasoning and neural networks of AI, resource estimates and decision support can be drastically enhanced.

Emotion aware technologies possess mood tracing capabilities with the potential to track and analyze an individual's emotional state (Morsy, 2016). This can enable construction professionals to keep mood journals and emotional data to understand what influences their emotion (Morsy, 2016). Additionally, authentication technology such as radio-frequency identification (RFID); quick-response codes; and fingerprint, vein, iris and facial recognition can allow for manpower management, credential verification and access control (Chin, Kim, & Choi, 2017).

Auto-Analytics

Auto-analytics and cognitive computing can be used to radically improve emotional intelligence and social relationships within the future construction workforce (Hansen, 2017). It is the process of intelligent logical analysis through technology, in which data is collected autonomously and used for analysis (RGA, 2013). Data can be collected through wearables such as watches, headbands and rings that can detect physiological data, such as heart rate, skin temperature and brain waves to accurately read a person's mood and provide real-time data to the user (ABI Research, 2017)

The Case of Australian Civil Construction

Construction has become one of the most important industries in Australia (Leviäkangas, 2017), and by 2020, Australia will become one of seven key countries that will account for 65% of the growth in global construction (Richardson, 2017). Using the case of a Sydney-based contract civil engineering firm specialized in design and asset management of transport infrastructure, land development and renewable energy projects, exploratory interviews were conducted to first unearth challenges faced by construction professionals in meeting leadership demands and expectations and then to explore the feasibility of existing digital technologies to assist in overcoming these challenges. The interviewees were a senior project manager with 16 years' experience, a project manager with 5 years' experience and a junior project manager with 2 years' experience.

Challenges in Addressing Leadership Demands in Civil Construction Projects

Interviewees expressed the significance of project leadership and its importance for the well-being of construction projects. They considered leadership as a critical component of their role and when asked whether they perceive themselves as a leader, one interviewee proclaimed "only because I have to be". Leading was considered as a way of being "consistent" and not allowing life experiences to interfere with project team experiences. The more junior respondents emphasized their thoughts on how leadership "comes with experience". Yet, the senior respondent did not entirely feel he had gained project leadership mastery or even gained a superior grasp of project leadership behavior. Rather, this respondent felt it was an ongoing challenge of his role.

The following sections outline the key challenges extracted from the interviews. They were clustered around the previously identified focus areas of leadership demands "team focus", "leader focus" and "project focus" (see Table 15.2).

Challenges Relating to the Team Focus

Respondents emphasized the necessity of cohesion and its importance to project success, highlighting that a difficult project with a "good team" can be successful, yet an easy project with a "poor team" may perform terribly. Interviewees felt that the difficulty in maintaining cohesion stemmed from the constant "chopping and changing" of project team personnel, which made it difficult to align team goals. Respondents also felt that leadership elements of team chemistry and cohesion were dependent upon "the people". As described, there are "different people" and "no one's good, no one's bad". It's all about "who they are, where they came from, and what their strengths and weaknesses are". This creates difficulties in dealing with the different types of personas, and aligning the way they operate with each other, with the best interests of the project. However, difficulties associated with cohesion were often related to difficulties associated with lack of control. With one respondent suggesting that "you don't necessarily build the team, you get given the team".

Respondents felt that the key challenges associated with communication arose from the need to communicate differently, with different types of people, and in different types of situations. As one respondent proclaimed; "it's not always rainbows and unicorns", at times you can "hurt people's feelings". This was further reinforced as interviewees lamented challenges with communicating with team members in the ways they expect to (or would like to) receive information. Getting the desired outcome from communicating was also seen as a pressing difficulty, with one respondent pointing out the difference in communicating, and "communicating effectively". "It's easy to tell someone to dig a hole, but it's more difficult to convince someone why they should be digging a hole". The challenge in communication is convincing people the reasoning behind what they are doing and "how it impacts the bigger picture".

Challenges Relating to Leader Focus

Respondents felt a key challenge in leadership awareness was in understanding people and how they operate. They believed that understanding which leadership style to adopt “takes up so much of your time” and if you get it wrong, “it can come back and bite you”. Another interviewee raised the issue of proactivity, and the difficulty in being proactive in an environment that often imposes reactivity. Getting appropriate feedback was also deemed challenging as respondents felt they could never really judge how others perceived them or their practices. Furthermore, respondents felt that maintaining a consistent persona was a challenging feat, as personal issues could change their behaviors and often result in poor leadership practice that could have “spill-over” effects throughout the project. Respondents articulated that; “when you are at home, you are not always expected to be consistent with your leadership”. But at work, leadership is a full-time commitment, and regardless of your situation, you still must be “calm”, give “direction” and be able to “think” clearly.

Interviewees revealed the notion of fluctuating control. They felt that at times there were inconsistencies with regards to their control over and control within the project, with one junior respondent describing how it can often be confusing “who is in charge”. The more senior interviewee emphasized how the lack of control can be “frustrating at times”, and how “you just need to manage” working within the limits of your authority, whereas the junior project manager articulated the notion of leading “upward”. Describing how leading also incorporates leading “your bosses” and the difficulties that arise when “convincing them to do things that you need”. This respondent felt that you could only lead as much as you could and ultimately the final “decision can be out of your hands”.

Challenges Relating to Project Focus

Respondents believed that operating as a leader within the civil construction industry was much more difficult than it would be in any other industry. One respondent highlighted the construction industry’s culture of “resistance”. He felt that there was a constant “resistance to change and resistance to do things differently” within the sector. Respondents also felt that there was a culture of “responsibility avoidance”, in which industry professionals would constantly disassociate themselves from any levels of ownership or autonomy. Thus, making it much more difficult to lead, as individuals would often refuse to take ownership of their work. One respondent also touched on the topic of health and safety appreciation and described how the civil construction industry is not "mature enough". This respondent felt that leading a safer environment was difficult due to the resistance from workers in which they perceive health and well-being initiatives "as making their job harder".

The overall consensus was that a project can significantly dictate leadership behavior. Respondents highlighted that the project setting meant that they were forced to "assess people quickly" and make "on the spot" decisions and judgments. They felt short projects imposed "minimal time upfront" to plan and implement leadership structure. Therefore, they lamented on "time" as an influential factor in leadership practice. There were many leadership activities that they wanted to adopt but felt they were unable to do so due to time and budget constraints. The emphasis from respondents was that they were not afforded the freedom to properly indulge in leadership practice, but given the time, they would have the opportunity to lead.

Table 15.3 summarizes the identified key challenges to meet leadership demands.

Table 15.3 Overview of challenges to meet leadership demands

Challenges relating to	
Team focus	• Effectively communicating in ways people expect to, or would like to receive information • Convincing members the reasoning behind what they are doing and how it impacts the bigger picture • Communicating differently, with different types of people, and in different types of situations • Constant changing of project team members • Understanding differences in styles, strengths and weaknesses • Difficulties with influencing the team you are given
Leader focus	• Understanding people and how they operate • Getting appropriate feedback • Pressures of maintaining consistency with leadership, even in times of difficult personal circumstances • Limits of authority • Final decisions are at times out of your control • Fluctuating levels of control can make authority ambiguous
Project focus	• Resistance of construction professionals to change • Responsibility avoidance among construction professionals • Lack of worker appreciation for health and well-being • Planning and thinking ahead with project time constraints • Budget doesn't allow for major leadership implementation • Project environment means you need to assess people very quickly

Potential for Digital Technology to Address the Leadership Challenges

In the second part of the exploratory interviews, the previous results were used to discuss how the capabilities of digital technologies could help overcome leadership challenges in the construction industry.

Respondents were able to articulate general and often vague methods they used in dealing with leadership challenges, ranging from "being consistent" to separating "work life from personal life". In short, the consensus was that "you just have to deal with it" as it comes. However, the junior respondent did suggest the use of the smartphone application "FaceTime" as a remedy to some communication difficulties faced. According to this interviewee, the lack of face-to-face communications was prominent in his position and made it difficult for him to effectively communicate. Therefore, the use of FaceTime allowed him to overcome these challenges and "bridge the gap" between less meaningful and meaningful communications. Respondents also felt that effectively communicating goals to all significant stakeholders could allow leaders to gain more authority within their positions. Furthermore, respondents believed that a great project leader could expand the objectives of the project itself beyond time, budget and quality, to involve "building" a "team" to "higher levels" of achievement in which everyone has "learnt", "grown" and "gained something".

Although the senior respondent was somewhat skeptical, the two more junior respondents could see immediate potential for digital technology to enhance leadership practice in some areas. One respondent made immediate connections with visual technologies, such as augmented reality, artificial reality, holographic display, as well as 3D, 4D and 5D BIM. He felt that these technologies would allow leaders to "sell the story" and help explain the reasoning behind objectives and goals. Moreover, this respondent felt that these technologies would be beneficial for leadership to "be able to show someone rather than tell them". "Having the capability of showing someone, having them see exactly what you are talking about, would help to sell that story". Respondents also felt that if the capabilities of auto-analytics were available, they would "use it tomorrow". The respondents were particularly excited about the capabilities to gauge factors like mood, sleep, physical and mental performance, as well as to provide real-time feedback on positive and negative words used in conversion, thus enhancing their leadership behavior. Furthermore, respondents believed that wearable technologies could assist with health and safety and would aid in a leader's responsibility to maintain workplace well-being and

Table 15.4 Digital technologies to assist with leadership challenges

Technology	*Capability*	*TF*	*LF*	*PF*
Smartphones	Can allow leaders to have a single mode of analysis for the project team, the project and themselves.	•	•	•
Wearables	Can allow leaders to maintain health and well-being of workers and positively influence organizational culture.	•		•
Drones	Can allow leaders to better track and report progress to team members, keeping them constantly informed and aware of their influence.	•	•	•
AR, VR & BIM	Can allow leaders to "show" rather than "tell", making it easier to "sell" the project story.		•	
Auto-analytics	Can allow leaders to understand themselves and their interactions with others.		•	

TF = team focus; LF = leader focus; PF = project focus

a positive environment. Respondents felt that utilizing wearables to build a cohesive team could influence the organizational culture, in which other organizations would "try to replicate" and "learn" from them, thus changing cultural perspectives within the industry.

However, all respondents argued that it would take more than technology to have an influence on the challenges outlined, and although they may assist with providing information, it would be up to the leader to use this information accordingly.

Table 15.4 outlines the key capabilities of digital technologies to assist with overcoming leadership challenges.

Discussion and Outlook

Digital transformation is fundamentally influencing all aspects of business and society. It can enable individuals and organizations to transcend from one way of working to another. In construction, emphasis is shifting from project management to project leadership, as professionals need assisting tools to not only aid in managing tasks and activities, but to aid in leading people.

Using the example of the Australian civil construction industry, we reviewed the existing body of knowledge on both industry-specific leadership demands and relevant digital technology capabilities to identify areas of improvement and to guide interviews with construction project managers. Findings indicate that digital technologies will have a positive influence on project leadership challenges. We find that project managers of civil contractors within Sydney understand the significance of leadership, but are often overwhelmed by its complexity. They demonstrate a willingness to adopt digital technologies as assisting mechanisms for leadership. Participants also lament on leadership challenges relating to project scope, industry attitudes and culture, level of authority, leadership habits/behavior, communication and team development. Construction professionals demonstrated specific examples in which they could utilize certain technologies to address challenges in their leadership practice.

Digital technologies can act as the catalyst that disrupts the behavior of a project team, which ultimately influences the environment in which the team operates. The input of digital technology within the leadership system can therefore transform leadership challenges into opportunities. As such, the adoption of digital technologies may hold the key in taking project leadership practice into the future. This chapter aims at opening up a discussion about exploring areas in which disruptive technologies can be used to create benefits beyond their primary technical applications.

References

ABI Research. (2017, December). *Wearable data analytics bring humans into the IoT*. Retrieved from ABI Research: https://www.abiresearch.com/press/wearable-data-analytics-bring-humans-iot/

Ahmed, M., & Ahmad, S. (2014). Transformation of smart phone to super phone: A future oriented gadget. *Pranjana, 17*(2), 1.

Akhavian, R., & Behzadan, A. (2016). Smartphone-based construction workers' activity recognition and classification. *Automation in Construction, 71*(1), 198–209.

Ameh, O., & Odusami, K. (2014). The leadership profile of Nigerian construction project managers. *Scientia Iranica: Transactions A Civil Engineering, 21*(4), 1241–1248.

Anantatmula, V. S. (2010). Project manager leadership role in improving project performance. *Engineering Management Journal, 22*(1), 13–22.

Arain, F. M. (2012). The quadrilateral model of leadership: Findings from a study on a mega project. *International Journal of Construction Project Management, 4*(2), 125.

Azhar, S., Jackson, A., & Sattineni, A. (2015). Construction apps: A critical review and analysis. In *ISARC. Proceedings of the 32nd International Symposium on Automation and Robotics in Construction* (p. 1).

Barber, E., & Warn, J. (2005). Leadership in project management: From firefighter to firelighter. *Management Decision, 7*(8), 1032–1039.

Blanco, J., Fuchs, S., Parsons, M., & Ribeirinho, M. (2018, April). *Artificial intelligence: Construction technology's next frontier*. Retrieved from McKinsey & Company: https://www.mckinsey.com/industries/capital-projects-and-infrastructure/our-insights/artificial-intelligence-construction-technologys-next-frontier

Cavaleri, S., & Reed, F. (2008). Leading dynamically complex projects. *International Journal of Managing Projects in Business, 1*(1), 71–87.

Chin, S., Kim, I., & Choi, C. (2017). What authentication technology should be chosen for construction manpower management? *Procedia Engineering, 196*, 309–314.

Clarke, N. (2012). Shared leadership in projects: A matter of substance over style. *Team Performance Management: An International Journal, 18*(3/4), 196–209.

Cullen, G. (2016). *The role of technology in health & safety*. Municipal World.

de Soto, B. G., & Adey, B. T. (2016). Preliminary resource-based estimates combining artificial intelligence approaches and traditional techniques. *Procedia Engineering, 164*, 261–268.

Del Rowe, S. (2017, September 29). *Digital transformation needs to happen now*. Retrieved from Destination CRM: https://www.destinationcrm.com/Articles/ReadArticle.aspx? ArticleID=120789

Dillow, C. (2016, September 13). *The construction industry is in love with drones*. Retrieved from Fortune: http://fortune.com/2016/09/13/commercial-drone-construction-industry/

Emil Berg, M., & Terje Karlsen, J. (2014). How project managers can encourage and develop positive emotions in project teams. *International Journal of Managing Projects in Business, 7*(3), 449–472.

Galvin, T., Gibbs, M., Sullivan, J., & Williams, C. (2014). Leadership competencies of project managers: An empirical study of emotional, intellectual, and managerial dimensions. *Journal of Economic Development, Management, IT, Finance, and Marketing, 6*(1), 35.

Gruman, G. (2016). What digital transformation really means. *InfoWorld, 18*(1), 1–3.

Hambling, D. (2015). A drone that learns. *NowScientist, 226*(3017), 20.

Hansen, S. (2017, May 25). *Analytics in the construction sector*. Retrieved from Sourceable: https://sourceable.net/analytics-construction-sector/

Heembrock, M. (2015). The risks of wearable tech in the workplace. *Risk Management, 62*(1), 10.

Heiskanen, A. (2016, July 27). *Augmented and mixed reality in construction*. Retrieved from AEC Business: https://aec-business.com/augmented-mixed-reality-construction/

Joan, D. (2015). Enhancing education through mobile augmented reality. *Journal of Educational Technology, 11*(4), 8–14.

Kappagomtula, C. L. (2017). Overcoming challenges in leadership roles–managing large projects with multi or cross culture teams. *European Business Review, 29*(5), 572–583.

Kassem, M., Benomran, L., & Teizer, J. (2017). Virtual environments for safety learning in construction and engineering: Seeking evidence and identifying gaps for future research. *Visualisation in Engineering, 5*(1), 16.

Khan, S. S. (2015). Importance of transformational leadership in project success: A theoretical framework. *Актуальні проблеми економіки, 1*, 67–76.

Kim, K. P. (2016). Investigation of readiness for 4D and 5D BIM adoption in the Australian construction industry. *Management Review: An International Journal, 11*(2), 43.

Kivrak, S. A., Arslan, G., & Tuncan, M. (2015). Implementing augmented reality in construction projects. *Applied Mechanics & Materials, 719–720*, 197–201.

Klashanov, F. (2016). Artificial intelligence and organizing decision in construction. *Procedia Engineering, 165*, 1016–1020.

Kouzes, J. M., & Posner, B. Z. (2003). *The leadership practices inventory (LPI): Participant's workbook* (Vol. 47). John Wiley & Sons.

Lazányi, K., & Dóka, L. (2015). Leadership practices in the Hungarian construction industry. *Managerial Challenges of the Contemporary Society. Proceedings, 8*(2), 55.

Lee, W., Kang, S., Moh, R., Wu, R., Hsieh, H., & Shu, Z. (2014). Application of BIM coordination technology to HSR Changhua station. *Visualization in Engineering, 3*(5). Retrieved from https://viejournal.springeropen.com/articles/10.1186/s40327-014-0008-9

Leviäkangas, P. P. (2017). Keeping up with the pace of digitization: The case of the Australian construction industry. *Technology in Society, 50*, 33–43.

Linder, C. (2017, June). *Wearable technology could save lives and dollars in construction industry.* Retrieved from Pittsburgh Post-Gazette http://www.post-gazette.com/business/tech-news/2017/06/12/Wearable-technology-construction-industry/stories/201706040035

Moon, S., & Seo, J. (2015). Integration of smart glass technology for information exchange at construction sites. *ISARC. Proceedings of the 32nd International Symposium on Automation and Robotics in Construction*, p. 1. Vilnius Gediminas Technical University, Department of Construction Economics & Property.

Moreno, C. (2014, June 21). *Free CAD software for mobile devices.* Retrieved from Cadalyst: http://www.cadalyst.com/cad/free-cad-software-mobile-devices-20019

Morsy, A. (2016). Emotional matters: Innovative software brings emotional intelligence to our digital devices. *IEEE Pulse, 7*(6), 38–41.

Müller, R., & Turner, R. (2010). Attitudes and leadership competences for project success. *Baltic Journal of Management, 5*(3), 307–329.

Nagy, P., & Koles, B. (2014). The digital transformation of human identity: Towards a conceptual model of virtual identity in virtual worlds. *Convergence, 20*(3), 276–292.

Oke, A. (2010). An examination of project management leadership styles of Nigerian quantity surveyors. *Journal of Building Performance, 1*(1), 57–63.

Park, J. K. (2016). Framework of automated construction-safety monitoring using cloud-enabled BIM and BLE mobile tracking sensors. *Journal of Construction Engineering and Management, 143*(2). Retrieved from https://www.academia.edu/35328870/Framework_of_Automated_Construction-Safety_Monitoring_Using_Cloud-Enabled_BIM_and_BLE_Mobile_Tracking_Sensors

Patterson, J. (2010). Leadership: The project management essential. *Production and Inventory Management Journal, 46*(2), 73.

Phelps, K. (2014). "So much technology, so little talent"? Skills for harnessing technology for leadership outcomes. *Journal of Leadership Studies, 8*(2), 51–56.

Pink, D. (2006). *A whole new mind: Why right-brainers will rule the future.* Penguin.

Pryke, S. L. (2015). The effect of leader emotional intelligence on leader–follower chemistry: A study of construction project managers. *Construction Management and Economics, 33*(8), 603–624.

RGA—Reinsurance Group of America. (2013, October 31). *Auto-analytics—Quantifying the self.* Retrieved from RGA—Reinsurance Group of America https://www.rgare.com/knowledge-center/media/articles/auto-analytics%2D%2D-quantifying-the-self#

Richardson, C. (2017). Technology offers the construction industry a brighter future. *Construction Engineering Australia, 3*(1), 16.

Sampaio, A., & Viana, L. (2014). Virtual Reality technology used as a learning tool in civil engineering training. In *Human System Interactions (HSI), 2014 7th International Conference* (pp. 156–161). IEEE.

Schuhmann, R. (2010). Engineering leadership education—The search for definition and a curricular approach. *Journal of STEM Education: Innovations & Research, 11*, 61–69.

Senam, M., Rashid, K., Sarkawi, A., & Zaini, R. (2014). Construction project leadership from the perspective of Islam. *International Journal of Islamic Thoughts, 6*, 46.

Senaratne, S., & Samaraweera, A. (2015). Construction project leadership across the team development process. *Built Environment Project and Asset Management, 5*(1), 69–88.

Slattery, D., & Sumner, M. (2011). Leadership characteristics of rising stars in construction project management. *International Journal of Construction Education and Research, 7*(3), 159–174.

Succar, B. (2009). Building information modelling framework: A research and delivery foundation for industry stakeholders. *Automation in Construction, 18*(3), 357–375.

Toor, S., & Ofori, G. (2008). Taking leadership research into future: A review of empirical studies and new directions for research. *Engineering, Construction and Architectural Management, 15*(4), 352–371.

Tyssen, A., Wald, A., & Spieth, P. (2014). The challenge of transactional and transformational leadership in projects. *International Journal of Project Management, 32*(3), 365–375.

Wright, R., & Keith, L. (2014). Wearable technology: If the tech fits, wear it. *Journal of Electronic Resources in Medical Libraries, 11*(4), 204–216.

Yngvason, Y. R. (2013). Jesus Christ as a project leader. *Procedia-Social and Behavioral Sciences, 74*, 398–407.

Zhang, L., & Fan, W. (2013). Improving performance of construction projects: A project manager's emotional intelligence approach. *Engineering, Construction and Architectural Management, 20*(2), 195–207.

CHAPTER 16

The Making of Data-Driven Sustainable Smart City Communities in Holiday Destinations

Sudipta Kiran Sarkar, Michalis Toanoglou, and Babu George

Introduction

Sustainable development is a term that emerged from the Brundtland Commission's report of 'Our Commons Future' in 1987 (Saarinen, 2006). The concept of sustainable development entails a responsible practice of resource consumption by current generations so that future ones are not deprived of the same resources for their basic survival (WCED, 1987; Saarinen, 2006). The argument in favour of advocating sustainable development got further impetus in institutional terms in the 'Earth Summit' in 1992 held in Rio de Janeiro, Brazil since the Rio Conference

S. K. Sarkar (✉) • M. Toanoglou
Woosong University, Daejeon, South Korea
e-mail: mtoanoglou@sis.ac.kr

B. George
Fort Hays State University, Hays, KS, USA
e-mail: bpgeorge@fhsu.edu

B. George, J. Paul (eds.), *Digital Transformation in Business and Society*, https://doi.org/10.1007/978-3-030-08277-2_16

in 1992 on account of the observation of manifold negative ecological and social impacts caused by decades of industrialization (Hanharan, 2010; Saarinen, 2006).

The perspectives of sustainability in tourism emanated from the Agenda 21, an action plan on sustainable development that emerged from the discourses of the Rio Conference. Sustainable development issues have become crucial in current tourism discourses (Roberts & Tribe, 2008)—given the negative impacts of tourism operations in terms of social, political, economic and environmental aspects. According to United Nations World Tourism Organization (UNWTO) 'Sustainability principles refer to the environmental, economic and socio-cultural aspects of tourism development, and a suitable balance must be established between these three dimensions to guarantee its long-term sustainability' (UNWTO, 2004).

Since the last three decades, several initiatives primarily by non-profit organizations as well as academic, governmental and inter-governmental organizations have pursued sustainability, both in practice and dialogue through regional, inter-regional and global projects, grassroots-based activism as well as workshops and conventions. In addition, private sector/industry initiatives towards sustainability have also been evident in spite of the conflicting interests emanating from a dominant neo-liberal economic environment that has shaped the industry perspectives in tourism. However, climate change issues and inequality in tourism destinations have been ever increasing despite various pursuits of sustainable development by different sectors in the tourism industry. Hence, the quest for sustainability in tourism is far from over due to the increasing environmental and social challenges faced globally.

Technological developments and innovations have brought us to a new phase in our modern society. Cutting edge technologies in digital, environmental and transportation domains have paved the way towards a new epoch of industrialization that provides accessibility, communication and socialization to the masses via affordable means. Developments in more efficient, faster and safe mass public transit systems, as well as developments in green energy that have made it relatively more affordable and cost-effective to a much larger section of consumers and businesses, have brought sustainability and technology closer and compatible with each other than ever before. Clean energy and cleaner modes of transportation driven by new innovative technologies and coupled with smart digital technologies have brought new opportunities to the process of transformation towards a sustainable future. Clean energy, cleaner modes of transportation systems as well as smart digital systems have fostered the development of sustainable urban tourism by virtue of hospitality outlets and urban attractions increasingly adopting green energy technologies—clean modes of transportation providing affordable, environmentally clean and efficient mobility services to tourists and digital technologies (powered by widely available Wi-Fi services) enabling communication and availability of information critical for destination knowledge and travel decision-making. Public/state funded technological projects around the world are leading the way for more innovations in cleaner automobiles and cleaner energy production—ARPA-E in the United States and Germany's KfW, a state investment bank have funded a number of clean technology initiatives with relation to energy efficient and renewable projects in the recent past (Tate, 2017). Tesla, a leading producer of electric cars in the United States has been funded by the US Department of Energy for producing electric automobiles with cutting edge technology. Returns from green technology investments initiated both by state and the private sector has resulted in creation of wealth for the public providing more coherence and impetus to the role of technology in fostering sustainable development (Harris, 2017).

The bond between sustainability and technology will mostly be necessitated by the imperative needs in cities and urban areas of the world. Growing complexities in ever expanding cities of the world resulting from the demands for quality of life, environmental protection and competitiveness will drive the need for sustainable development powered by clean energy technology, soft mobility technology and more intelligent digital technologies (Siemens AG, 2013). Moreover, cities and other urban areas around the world, which currently produce approximately 70% of the greenhouse gas emissions, will increasingly become the epicentres of high volumes of consumption leading them to be major contributors to climate change (Siemens AG, 2013). Hence, the demand for innovative technologies that can facilitate sustainable and inclusive urban growth will necessitate effective solutions and strategies that can ensure resilient and environmentally compatible ways of urbanization of our cities in the future.

One of the prime areas of economic activity and consumption in cities around the world is tourism. Cities and urban areas vibrant with arts, culture, music, architecture and design and heritage along with centres generating robust volumes of business and commercial activity have attracted all kinds of visitors. Moreover, ease in mobility and communication due to rapid digitalization and internet connectivity via Wi-Fi as well as expansion of different modes accessible and efficient public transportation networks have made tourism in urban areas grow manifold. Urban tourism as referred to by UNWTO (2012) is 'trips taken by travelers to cities or places of high population density'. Cities which are administrative capitals of nations are gateways to many destinations and the first point of reference for tourists in acquiring an impression of the nation as a destination. This makes large number of tourists travel to and via cities and hence crucial for destination management organizations to manage cities as destinations in a way that they can facilitate quality experiences and enhance satisfaction of such large mass of tourists.

This chapter discusses the magnitude of technology being strongly implanted in the process of sustainable development of cities as manifested through urban tourism—one of the main pillars of urban economy. The chapter identifies four critical dimensions of sustainability in urban tourism—urban ecotourism, food tourism, urban culturescapes and smart transportation. Correspondingly, critical dimensions of technology in urban tourism—social media and user-generated content, gamification, M-learning (mobile-based learning), green energy technologies and green/soft mobility, are also discussed comprehensively. The chapter then discourses the integration between technology and sustainability in urban tourism contexts based on existing cases and models around the world.

Dimensions of Sustainability in Urban Tourism

Culture, nature and communities in urban areas have been instrumental in shaping the tourism landscape of cities around the world. City-based parks, gardens and waterbodies as well as museums, arts galleries, performing arts, avant-garde arts and countercultural communities have attracted large number of culturally inclined independent travellers. On the rise are emerging forms of urban travel activities like food tourism combined with walking tourism that involves trips to areas of cities noted for local food—wet markets, restaurants, confectionaries and all other forms of eateries and food outlets. Also associated with such tours are the communities engaged with food tourism businesses, their enterprises, the aesthetic aspects of their culinary offerings, their skills as well as the socio-cultural environment they create and belong to as a whole. Another aspect that culture-oriented city walking tours involve are visitations to prime natural and cultural attractions of urban areas harnessing mass transit public transport systems. All these elements reflect socio-cultural inclusivity, environmental compatibility and offer deeply engaging lived as well as socializing experiences to tourists. Such increasingly popular forms of urban tourism have been receiving adequate patronage from tourists and destination planners in the recent times as the modern tourism discourses progressively reflect the virtues and the essentiality of sustainability.

Urban Ecotourism

Towns and cities have large tracts of nature either in the form of national parks, sanctuaries, wildlife reserves and other organic forms or in modified or artificially created forms. Urban centres around the world are increasingly trying to harness the value of such nature tracts in providing sustainable living to both its residents and visitors. Urban tourism now significantly involves travel to nature-based areas in and around cities especially to such natural areas, which are of high significance in terms of flora or fauna or both. Such trips help both residents and visitors understand the value of nature as well as enable them to have a sustainable approach to experience cities. Such a phenomenon can be referred to as 'urban ecotourism'. Though the term 'urban ecotourism' was first coined in 1996, the phenomenon has been in practice since long. Planeta.com, an award-winning website on ecotourism, in a conference in 2004, described it as 'simply nature travel and conservation in a city environment' (Kastelein, 2004). The definition laid down by the Toronto Green Tourism Association (2006) was 'travel and exploration in and around a city that provides visitors and residents with a greater appreciation of the cities' natural resources and cultural resources'.

Cities like New York, Toronto, Singapore, Hong Kong and Kuala Lumpur have attracted local and international visitors towards both natural and modified nature-based attractions. Nature-based attractions in these cities offer opportunities for various leisure and adventure activities like bird-watching, nature-viewing, photography, caving, hiking and rock-climbing. Such opportunities have made nature-based attractions in these cities an integral part of their urban tourism offerings to all kinds of visitors. One of the prime features of urban nature-based attractions in most cases is that they are often well-connected by mass transit public transportation systems due to their location within or on the periphery of cities and major towns. Hence, travel to urban nature-based attractions or urban ecotourism contributes towards sustainable development as it creates opportunities for alternative recreational activities as against the traditional highly consumptive urban tourism activities. Urban ecotourism apart from primarily focusing on travel to nature-based areas in urban areas also involves trips to cultural sites like museums, art galleries, handicraft streets, wet markets and food streets. This provides the social and anthropological element to the urban ecotourism phenomenon making it inclusive and environmentally inclined at the same time. Urban ecotourism offers the opportunity to all sections of the society to be educated on the value of ecology to human societies as nature-based attractions offers easy access by the virtue of being located within or close to cities and by being well-connected efficient and affordable mass transit transport systems.

Food Tourism

Food tourism is the phenomenon of travelling and experiencing other cultures through food (Herrera, Herranz, & Arilla, 2012). Food tourism is also sometimes referred to as culinary tourism or gastronomique tourism. Food tourism entails a sense of place in geography (or the link between land and people). Travel for the purpose of experiencing exquisite culinary offerings of a destination has become one of the main motivators of travellers in terms of their travel behaviour and activities at a destination (Herrera et al., 2012; OECD, 2012). Such culinary experiences are offered in an array of settings—the most exciting and engaging ones being local street food at popular street food enclaves of towns and cities, wet markets or night markets as well as local/traditional food outlets which by virtue of their unique offerings (taste, aroma and aesthetic style of presentation) for a considerable period of time have become transformed into a heritage due to the sustained popularity that they have enjoyed (Gillan, 2014; Nualkhair, 2015). Experiencing food related events like food festivals or experiencing hands-on cooking (participating in cooking classes) as well as food tasting and/or experiencing the attributes of specialist food production are also some of unique experiences that urban destinations offer to foodie travellers (Peltier, 2015).

Apart from the hedonic and consumptive aspects associated with the experiences of food tourism, the communal and cultural elements that are strongly embedded in it often construct the basis of such experiences (Peltier, 2017). A critical aspect associated with these communal and cultural elements is sustainability. Food tours are often characterized by walking tours for short durations of 4–5 hours as well as harnessing the services of mass transit public transportation systems. The use of such soft modes of mobility enable food tours to deeply involve the communal element as it provides a social interface between foodie travellers and the communities engaged with food tourism businesses, the small scale nature of the enterprises they run as well as the aesthetic aspects of their culinary offerings along with their culinary skills (Gillan, 2014; Nualkhair, 2015). This social interface constructs a socio-cultural environment that immerses the visitors and their gazes with deeper social meanings of their experiences of local culinary cultures. Hence, the utilization of soft mobility and the socially immersive experience reflect the sustainable nature of the communal and cultural elements of food tourism.

Urban Culturescapes

Cultural activities ranging from visitations to art galleries and museums depicting different aspects of aesthetic and social scenes of modern society as well as radical/avant-garde and contemporary forms of arts and events mainly advocated by the urban youth shape the culturescapes of our cities (Merrick, 2011; Buffenstein, 2016). Urban areas around the world encompass both mainstream and alternative forms of visual and performing arts along with edifices of heritage like museums and monuments that attract a large number of travellers to urban areas.

Cities like Berlin, London and New York have attracted culturally inclined travellers to art galleries and art collectives that focus both on popular and 'avant-garde' forms of aesthetic expressions. Attractions like the Museum island in Berlin which consist of museums and galleries as well as alternative art centres like Kreuzberg, Friedrichshan and Prenzauler Berg attract all kinds of politically and art-inclined visitors. Many of such culture streets represent underground and alternative urban youth cultural expressions in the form of street musicians and street art collectives. Cultural centres like these in many instances have been the backdrop of radical urban socio-political movements that gave rise to countercultural thoughts challenging traditional Capitalist consumption models and advocating collective action and Do-It-Yourself ethos. Such actions and ethos have become underlying principles of many urban sustainable projects. Urban cultural centres like these often undergo continuous evolution and in the process have attracted a wide array of visitors—art-inclined people and artists themselves, youth belonging to avant-garde movements as well as with visitors based on socio-political factors like immigrants and anarchists. Urban culturescapes that involve such vibrant youth cultural vibes offer visitors the experience and learning on ecological and social dimensions of urban sustainability through different artistic expressions and other culturally expressive activities.

As in the case of urban ecotourism and food tourism, visitations to urban landscapes also involve the harnessing of the urban mass transit public transport system that connect major centres of cultural sites with various important mass transit hubs.

Soft Mobility

Soft mobility refers to low emission transportation systems that provide mobility in urban spaces to the masses. Such low emission transportation systems operate in a variety of modes like inter-modal, waterborne, rail, public/collective transport and non-motorized individual mobility (Rocca, 2009). Soft mobility should involve—the consideration of the urban destination geographical features, legal and planning structures anticipated ecological, social and financial impacts of development as well as the needs and wants of the residents of the urban destination (Kramer, 2009; Transdanube, 2014).

Cycling/biking in urban streets has become a popular phenomenon around the world. Tourists and residents alike commute considerable distances in their daily schedules in cities with the help of cycling/biking. Several urban destinations have become bike-friendly cities encouraging visitors to utilize cycling paths as a way to reduce congestion and automobile-based pollution. Cheap and affordable mass transit systems like subway trains, monorails, trams and electric buses have provided easy accessibility to different cultural, nature-based and gastronomique streets to all types of visitors (Rocca, 2009). Therefore, one of the emphasizing factors emerging from soft mobility is to convert more tourist movement into sustainable modes of transport which can also be referred to as modal shift (Gebhardt et al., 2016). An inter-modal chain model of mobility for tourists involving the combination of waterborne, rail and public/collective transport as well as walking and cycling ensures a sustainable consumption from tourism movement within an urban destination (Gebhardt et al., 2016).

The rise of collective transport or transport options that have emerged under the sharing economy has also reduced the rate of consumption and offered a more convenient, reliable and safe mobility options for tourists in cities and towns. Sharing economy in transport refers to carpooling services where car journeys are shared by more than one person. Several carpooling initiatives like Lyft, MyCotra and Uber have become commercially successful and are contributing towards less congestion and reducing individuals' travel costs.

In the Asian context, major cities like Singapore, Hong Kong, Seoul and Tokyo offer a combination of transport options ranging from subways or referred to as MTR and MRT (in Hong Kong and Singapore respectively) supported by extensive citywide bus networks along with options for biking-cycling at different parts wherever possible.

The Case of Hong Kong as a Soft Mobility Destination

In Hong Kong, the MTR (Mass Transit Railway), a combination of heavy rail, light rail and feeder bus service, is one of most extensive and profitable mass transit systems in the world and offers itself as a successful model for many other mass transit systems coming up around in the world and particularly within Asia. In addition, Hong Kong offers the option of using tram services in certain parts of the highly congested Hong Kong Island as well as extensive cycling opportunities in the new territories. Besides, waterways—a traditional mode of transport that have existed in Hong Kong since long still operates significantly offering services from Hong Kong Island (Central) and Kowloon ferry terminals to different parts of Kowloon, Hong Kong Island, New Territories and several outlying islands. The tram and ferry services are also a major part of the tourist experiences of Hong Kong. Tourist—centric transport like the furnicular rail or the Peak tram which covers a distance of 1.4 kilometres approx. starting from downtown Hong Kong Island to the Victoria Peak, a point of commercial and tourist importance at the higher elevations of Hong Kong Island enables both tourists and local residents a convenient commuting option. Another tourist-centric transport is the Ngong Ping cable car which starts from the Tung Chung MTR station to the tourist attractions in the higher elevations of Lantau island (Lantau peak) covering a distance of 5.7 kilometres and providing breathtaking views of the Hong Kong International Airport, the dense flora and topography of Ngong Ping plateau as well as the urban areas around. The Hong Kong International Airport and the

(*continued*)

(continued)

Hong Kong Disneyland are connected with the main commercial and residential areas of Hong Kong by means of MTR's special services—the Airport Express and the Disneyland Resort Line respectively. The urban bus network operated by different companies as well as the Green Liquified Petroleum Gas (LPG) light buses (minibuses) offers a comprehensive and highly well-connected system of road transportation to residents and tourists alike. Road transport in Hong Kong is also featured by three kinds of LPG run taxicabs—the red, green and blue taxicabs. The red taxicabs operate in the downtown areas of Hong Kong Island and Kowloon, the green taxicabs operate in the New Territories of Hong Kong and the blue taxi cabs operating only in the Lantau Island.

This comprehensive mass public transport system of Hong Kong offers a perfect example of an inter-modal model of soft mobility in an urban tourism destination. Firstly, the MTR services, the ferry services and bus network (especially the green LPG minibus services) as well as the extensive cycling-biking options which are highly interconnected enable tourists to commute to different famous and offbeat folk cultural, local culinary and heritage attractions of Hong Kong. Secondly, the 'Peak Tram' and the Ngong Ping Cable car offer exciting and safe tourist-specific experiential commuting to a large of urban tourists and resident visitors. Thirdly, the engaging experiences tourists obtain while availing ferry services and the tram services to commute around Hong Kong transform such transport services into tourist attractions. The extensive MTR network, cycling-biking options as well as the extensively operating LPG run cabs and minibuses along with tourist-centric transport options of the 'Peak Tram' and the Ngong Ping Cable car aptly manifest an eco-friendly and inclusive mass-oriented soft urban mobility system. Due to this dynamic nature of the urban inter-modal in the city, Hong Kong has been ranked first in terms of having the most sustainable transport as per the 2017 Sustainable Cities Mobility Index by Arcadis (Arcadis, 2017).

Dimensions of Technology and Their Integration with Dimensions of Sustainability in Urban Tourism

Cities and urban regions around the world are the most technologically vibrant in terms of use and applications. Urban infrastructure provides the basis and urban consumption patterns necessitate the presence and use of technology at considerable levels. Cities are connected with Wi-Fi technologies and its commercial activities are highly dependent on the use of innovative tech applications for various interconnected systems to work efficiently. Urban regions around the world are generally equipped with technological connectivity in terms of Wi-Fi availability, internet zones and social media sharing that benefits visitors and tourists to derive satisfaction and intention to re-visit urban attractions.

Social Media and User-Generated Content

Social media founded on the technological applications of user-generated content (Web 2.0) plays a major role in communication and socialization among tourists (Buhalis & Law, 2008; Kaplan & Haenlein, 2010). Urban destinations facilitated by easy availability of Wi-Fi services drive online activity of tourists who prefer to be connected while they are in the process of consumption of urban leisure-based experiential products (Bock, 2015). Facebook, which enjoys the highest number of users in the world, followed by Twitter and travel-oriented platforms like Trip Advisor—have been the widely used social media platforms by tourists for online social interactions, sharing of vital travel information influencing travel decision-making by tourists (Santos, 2011). Online social exchanges and sharing of information by tourists in online platforms have led to co-creation of collective knowledge (Seraj & Ayesugul, 2012).

Eco-conscious tourists in urban settings have the ability of setting ecological citizenship through continuous co-creation of knowledge that is collectively generated through their online social interactions and sharing of information on ecological and social elements they experience in nature-areas in urban settings (Rokka & Moisander, 2009; Sarkar, Au, & Law, 2013). The advent of Flickr, a photo-sharing social networking site and recently, Instagram—a photo-sharing social networking app has initiated online co-production of destination knowledge through pictures and photographs. This has enabled tourists create aesthetically credible pictorial connotations of destinations that have deeper influences on fellow-tourists.

Urban tourism destination-based authorities and promotional bodies have harnessed the co-creative and collective knowledge building potential of social media platforms like Facebook and Instagram (Sarkar et al., 2013). Promotional efforts on social media platforms and initiating co-creative practices among consumers have become a key innovative marketing strategy that transforms consumers into pro-sumers. Co-created content by tourists on social media can become invaluable marketing content for destination management organizations (Özdemir & Çelebi, 2015).

The advent of smartphones in the last 5–7 years have enhanced the role of social media use among urban tourists. With the increasing availability of Wi-Fi facilities and high-speed internet connectivity in public places, cultural sites as well as natural sites, smartphones have accelerated the volume of online activity of tourists. Smartphones have drastically reduced the time between tourists experiencing an attraction and their sharing of pictorial, audio-visual as well as textual information in various social media platforms based on such experiences. Tourists can very quickly engage in co-creative practices on social media platforms in the context of urban destinations due to the widespread availability of Wi-Fi and increasing use of smartphones and smartphone-based apps. Tourist-generated (and co-created) content in social media platforms primarily via smartphones have become more reliable and easily available than supplier-generated content resulting in effective travel decision-making and collective knowledge building by existing and prospective tourists.

In the context of urban tourism, both tour operators and tourists make use of social media platforms like Facebook and Instagram to co-create experiences and products of nature, culture and gastronomical tourism. Moreover, blogs maintained by tourists on urban nature, food and cultural attractions play a major role in initiating collective knowledge building through sharing and exchange of key information. Such collective knowledge or tourist/consumer generated knowledge become more intense and reliable than supplier-generated knowledge as it involves a dialectic process of critical and comprehensive introspection of content. Moreover, through such dialectical process of collective knowledge building, tourists establish and advocate ecological and cultural citizenship online.

Gamification in Tourism

Gamification in tourism is 'the use of game mechanics and experience design to digitally engage and motivate people to achieve their goals' (Burke, 2014). Gamification can be a great method to digitally engage visitors with a use of their smartphones, tablets or other digital devices (Sever, Sever, & Kuhzady, 2015).

Gamified tourism experiences often involve challenges in the form of quizzes, puzzles, making and submitting photos according to a task. Scores based on these challenges determine how well the participant has accomplished the challenge (Kazakova, 2015). To monitor the participants there are leaderboards who keeps them updated how they are progressing in the game in terms of the rounds of challenges to be completed and the scores scored. Those completing all stages of the game successfully with high scores are rewarded in both virtual and real (tangible) forms (Kazakova, 2015).

Gamified tourism experiences are offered in different forms especially in the urban context. Some of the prominent forms are location-based augmented reality games, gamified tours specific to urban environments, augmented experiences of theme parks and transmedia storytelling, gamified immersive experiences in cultural and natural heritage and virtual cultural heritage experience, gamified restaurant and hotel experiences.

As mentioned earlier, one of the main aims to offer gamified experiences to tourists is to immerse them in a situation that is highly subjective and enduring in nature and therefore can provide high levels of satisfaction to them. Gamified experiences have the potential to bring loyalty among tourists for the destination and enable repeat visits. Gamified experiences are an attempt to add value to the generic trip experience of tourists while they are visiting an urban destination (Sever et al., 2015).

Applying gamified experiences in the area of sustainable tourism can facilitate social interactions, cultural understanding, inculcate sustainable consumption habits and bring about an enhanced quality of experience (Negruşa, Toader, Sofică, Tutunea, & Rus, 2015). Gamification in sustainable tourism is aimed at engaging as well as encouraging tourists to advocate and internalize new sustainable habits, sustainable consumption or even engage in new sustainable activities and visit new destinations that are sustainable. Gamified experience in the context of sustainable tourism has the potential to provide intrinsic incentives to tourists while they are engaging in activities that enable them to derive self-esteem and social recognition (Negruşa et al., 2015). The merits of applying gamification in sustainable tourism involve usher engagement before during and after any nature-based experience as well as marketing and promotional benefits. Gamification by means of improving the experience offers an opportunity to attract new visitors and new markets.

M-Learning or Mobile Learning

M-learning or mobile-based learning is a process of learning and education through the means of movable electronic computing devices like smartphones and tablets. The use of M-learning in the hospitality industry has been evident in the recent times in areas of hospitality workers' training and education. Hospitality chains like the Marriot and Hilton are applying M-learning applications for recruitment of new workforce in their respective hospitality properties.

The potential application of mobile learning in the context of the touristic use can be realized in the form of environmental interpretation and visitor education within the context of urban natural settings like botanical gardens, city parks, urban groves and other similar places. Utilizing the current developments and innovations in mobile broadband networks, smartphone technology and mobile software applications, the M-learning process can provide engaging interpretive programmes involving free-choice learning and mindful visitor experiences to tourists in urban natural settings.

Soft Mobility and Green Energy Technologies in Urban Tourism

As discussed earlier, soft mobility involves a set of different forms of low emission transport in urban destinations—an inter-model mode of transport. Soft mobility forms have facilitated accessibility to the masses and brought about social transformations and quality of life in many parts of the world. Inter-modal soft mobility has provided inclusivity and an egalitarian mode of transport in which lower income urban communities have found easy and convenient access to different parts of urban areas where they have to commute for earning a means of livelihood and education. Efficient and affordable inter-modal soft mobility available for local residents from all sections of the society also provide accessibility and connectivity to tourists and visitors in urban destinations. Such inter-modal soft mobility forms are driven by technologies that provide access to difficult terrains and dense settlements evident in urban destinations. Transport technologies like cable car and low emission automobiles using new forms of renewable fuel as well as electricity have provided greater social sustainability. These are the emerging smart mobility technologies that have enabled visitors from all economic sections or urban societies to affordably have access to attractions of ecological and socio-cultural interest.

The Case of Gondolas or Cable Cars in Latin American Cities and Hong Kong

Gondolas or cable cars are considered as a cheap option which requires lesser investment for infrastructure and lesser need of displacement of people and communities compared to other mass transit forms like subways or rail tracks. Gondolas may not have the ability alone to transport a large number of people from one to another but support significantly other forms of mass transit system by playing a major role in inter-modal connectivity (Barber, 2017). Gondolas on one hand can facilitate tourist movement within and to attractions in urban destinations and on the other hand also enable to move local residents from their neighbourhoods to major centres of cities at an affordable cost especially where the urban geographical landscape is characterized by challenging topographies. In particular, gondolas have been able to provide a faster, time-saving and affordable mode of transport to urban residents living in the poorer parts of cities and linking them with commercially important parts where there are opportunities for earning a livelihood. Latin American cities like La Paz (Bolivia), Medellin and Cali (Columbia) and Caracas (Venezuela) are major examples of such socially inclusive and effective gondola systems in commuting terms for the masses (Barber, 2017).

The Metrocable in Medellin city, Columbia is regarded as the world's first integrated multi-line gondola system. The Metrocable connects with major business and commercial centres of Medellin city with many outlying and less wealthy areas of the city in upper elevations of hillsides. The Metrocable was opened in 2004 and since then has been credited on one hand, in reducing commuting times, providing cheap and efficient connectivity to many low income residents of outlying areas and on the other hand, has been significantly instrumental in reducing crime rates and urban poverty levels in the city(Barber, 2017).

Mi Teleférico, the gondola system in La Paz, Bolivia is considered as one of the most developed and longest in the world. The Mi Teleférico was opened in 2014 and since then has enabled 50 million people to commute from poorer parts of the city to the upscale business and commercial areas. It has been reported to have saved 652 million minutes for all commuters it has served since 2014. The Mi Teleférico has been claimed to be a very effective project and the city authorities are considering to expand it to seven more lines which are expected to be covering a distance of 20 kilometres.

(*continued*)

(continued)

The Caracas Metrocable in Caracas, Venezuela operates with the similar philosophy of providing connectivity for the city's poorer parts with the affluent parts. It was built to avoid displacing people and providing poorer parts of Caracas like the San Agustin a safe and efficient mode of transport to commute especially within such areas. The Caracas Metrocable primarily connects with other major forms of public transport systems of Caracas and provides transportation within the San Agustin area of Caracas.

All the above examples manifest an environmentally significant and socially inclusive efficient form of urban soft mobility driven by the Cable car or gondola technology. Such a gondola system has proven to have provided safe and crime free transportation which becomes an important factor in tourist's use of such soft mobility systems.

The Ngong Ping 360 cable car in Hong Kong and the Singapore Gondola are on the other hand, urban projects that have been purpose-built for tourist mobility. These cable car projects have been able to divert tourist movement away from the vehicular traffic-prone areas of the city. Moreover, these projects have reduced the dependence of road transportation that involves automobiles run by fossil fuels especially in the case of Lantau Island, Hong Kong.

The Case of Electric Automobiles and Bio-Fuel-Driven Urban Buses

Another variety of smart mobility technologies are electric/hybrid vehicles exemplified in the form of electric cars and bio-fuel driven vehicles. These vehicles run on fuel from renewable sources like bio-fuels and electricity.

One of the recent and best examples of bio-fuel driven is the Bio-bus experimentations in the UK. Bio-bean, a London-based enterprise, has taken the initiative to run a project of Bio-buses driven by the bio-fuel made out by combining oil extracted from old coffee grounds and diesel (Hirtenstein, 2017). Utilizing old coffee grounds from major coffee shops across the UK, it is claimed that 6000 litres (1583 gallons) of bio-fuel or B20 bio-fuel can be produced to run a sizable fleet of buses for public transport in London city (Hirtenstein, 2017). Though bio-fuel is already used to run 9500 buses, the 'coffee-powered' bio-bus is an innovative soft mobility technology to be used for the first time in the city (BBC, 2017).

(*continued*)

(continued)

Another bio-fuel-driven bus project also from the UK is the Bio-Bus project also known as GENeco Bio-Bus. The Bio-Bus is a project powered by bio-fuel in the form of gas formed out of human wastes of food, sewage and commercial liquid wastes (Geneco, 2016). Though two Bio-Buses were operating in 2015, the Office for Low Emission Vehicles turned down the bid for funding and operating this project in 2016 (BBC, 2017). However, officials at GENeco—the enterprise which ran this project—believe the concept that emerged from this project can inspire other similar projects that can become successful since bio-methane gas powered buses have been found to reduce significant amount of air pollution in urban areas around the world (BBC, 2017).

Electric cars are now becoming popular in many countries around the world and in the United States, sales of electric cars rose to 37% in 2016 compared to the previous years and a 70% year-to-year increase in monthly sales (Rapier, 2017). This has made Hotel chains like the Marriot, Starwood and the Hilton to open Electric Vehicle (EV) charging stations in collaboration with Tesla (the largest electric automobile maker) and GE in many of their properties. This move by hotels to welcome electricity powered low emission vehicles also serves their strategic purpose of alluring high-end guests (Hanley, 2015). Such high-end guests who can afford the most expensive services offered by hotels also happen to have the ability to afford electric cars. Apart from hotels, car rental services are starting to offer electric car rentals in some parts of the world. The case of Enterprise Rent-A-Car electric car rental services in Orlando city exemplifies the use of electric cars in the travel industry. The electric car rental service named as Drive Electric Orlando (DEO) programme started in 2013 as a result of collaboration between Enterprise and Electrification Coalition and has grown considerably through the strength of tie-ups with hotels, theme parks and tourism promotional organizations like Visit Orlando (Padilla, 2017). The DEO programme offers opportunities to tourists for test drives of the latest versions of the electric car model—Chevy Volt, a hybrid electric car brand of Chevrolet (Hanley, 2015). Through such an opportunity, tourists are in turn provided the experience of new low emission vehicle technologies through car rental services and motivated to be owners of electric powered low emission vehicles in the near future (Padilla, 2017). The DEO programme is facilitated by 300 EV charging stations in different tourist centres of Orlando city and 40 partner hotels that provide free parking and valet services to the tourists who are clients of the program.

Green Energy Technologies

Green energy technologies are exemplified by the existence of energy conservation and cost-saving technologies primarily in the context of urban hotels. The casino destination of Las Vegas has demonstrated how major casino and convention facilities can be majorly dependent on renewable energy sources instead of conventional energy sources.

The hotel industry has also improvised its process of energy saving by installing digitally powered rooms and energy management systems that control and minimize energy consumption in guest rooms. Though this technology is not specific to urban hotels, its applicability has been possible more in such hotels. However, one of the best developments of green energy technology in the context of urban tourism has been in the case of Gardens by the Bay in Singapore.

Gardens by the Bay, Singapore

Gardens by the Bay opened in 2012 demonstrates a model of an iconic green urban attraction that involves the harnessing of a range of different renewable energy technologies like solar energy, biomass technologies and technologies providing the basis of sustainable design (Rowell, 2017). The most striking technological element of this attraction is the manmade supertrees. These supertrees ranging between 25–50 metres in height are built with a combination of green technology features—solar energy harvesting photovoltaic cells on the canopies as well as water harvesting inbuilt tanks that enable cooling of domes in the attraction containing 90,000 plants (Rowell, 2017). The supertrees themselves support 160,000 plants and the energy produced by the photovoltaic cells on their canopies are used for spectacular light, colour and sound displays from the attraction during nighttime (Rowell, 2017). The two domes or glass biomes in the attraction—the Flower dome and the Cloud Forest—are equipped also with sustainable features like green designing which allow sunlight to provide nourishment to the plants inside as well as rainwater capturing features that allow irrigation of the different flora species that they contain. The Gardens by the Bay attracts 25 million tourists every year from all over the world since it was opened in 2012 (Algie, 2014).

Gaming and entertainment companies in Las Vegas have moved away from conventional-energy-producing agencies like NV Energy. MGM Resorts International, one of the largest gaming and entertainment companies in Las Vegas exited from NV Energy paying an exit fee of $87 million (Hernandez, 2016). It is now considering a full-fledged transformation into a renewable energy agency in terms of fulfilling its energy needs by harnessing its phenomenal rooftop solar project as well as tie-ups with energy companies investing in solar energy as well as with solar photovoltaic developers and solar plants projects outside its premises (Hernandez, 2016). MGM developed rooftop solar projects in one of its principal Las Vegas projects—Mandalay Bay—which now has 26,000 solar panels spread across 28 acres and producing over 14 megawatts of energy. Other Las Vegas resorts like the Wynn Resorts and Las Vegas Sands are planning to buy more solar energy from the open market produced in surplus volumes in the solar farms in nearby places as well as produce renewable energy particularly solar energy themselves.

Conclusion

As the drive towards the enhancement of sustainable and smarter urban tourism grows (UNWTO, 2017), the connection between technology and sustainability will get further strengthened. On one hand, as the world moves towards the era of Industry 4.0, (technologically) smarter cities will receive more impetus entailing the growth of smart tourism destinations—driven by digital technologies, widespread availability of internet technology and smarter digital devices. On the other hand, as the climate crisis becomes increasingly significant, cities will have a larger role to play in the deployment of green tech and clean energy which will, in turn, become embedded in the functioning of major urban industries like tourism. Moreover, smart mobility options are being increasingly adopted by global cities in the world which focuses on offering efficient mass transit systems and clean energy/low emission automobile and inter-modal transport options to both tourists and local residents (Arcadis, 2017).

A summary diagram highlighting the interlinkages discussed in this chapter is presented in Fig. 16.1.

The World Economic Forum (WEF) considers innovation in technology to be the key for growth in the times to come principally in the areas of digitalization, smart systems (encompassing intelligent systems in clean energy, soft mobility and man-machine interfaces), artificial

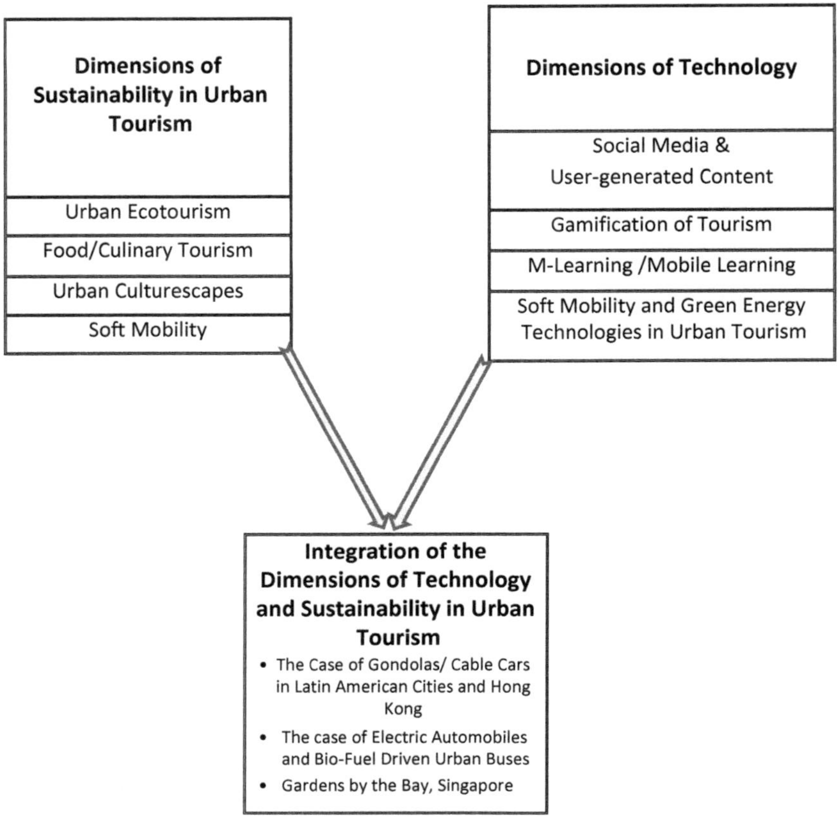

Fig. 16.1 Interlinkages between sustainability and technology in the urban tourism setting

intelligence and sustainable technologies (World Economic Forum, 2017). The UNWTO's call for a sustainable tourism for development in 2017 in the context encompasses social and economic inclusivity, cultural diversity, cultural mutualism and clean energy-driven environmental practice and (UNWTO, 2017). Therefore, the connection and integration of sustainability and technology in urban tourism contexts need to entail the social and cultural dimensions of both as their fundamental basis in addition to their environmental and innovational aspects.

The integration of sustainable forms of tourism and social and environmental dimensions of technology in urban contexts must yield benefits for all sections of visitors as well as all sections of the local resident population in urban destinations of the world.

One of the fundamental considerations for urban destination management organizations should be to put the sustainability-technology integration as the main underlying element in their destination marketing and branding process. Future studies focusing on the sustainability-technology integration may focus on the context of specific destinations—their scope in terms of sustainable forms of urban tourism attractions available as well as the available forms of digital and clean energy technology and their extent of use in such contexts.

References

Algie, J. (2014). Gardens by the Bay, Singapore: The coolest gardens in the world? Retrieved December 12, 2017, from https://jonistravelling.com/gardens-by-the-bay/.

Arcadis. (2017). *Sustainable Cities Mobility Index 2017: Bold moves*. Arcadis.

Barber, M (2017). 11 urban gondolas changing the way people move. Retrieved December 29, 2017, from https://www.curbed.com/2017/9/21/16340394/urban-gondolas-cable-cars-cities.

BBC. (2017). London buses to be powered by coffee. Retrieved December 21, 2017, from http://www.bbc.com/news/uk-england-london-42044852.

Bock, K. (2015). The changing nature of city tourism and its possible implications for the future of cities. *European Journal of Futures Research, 3*(20), 1–8.

Buffenstein, A. (2016). Icelandic music festival hosts impressive Avant-Garde Art Program. Retrieved December 13, 2017, from https://news.artnet.com/exhibitions/cycle-festival-2016-avant-garde-art-694697.

Buhalis, D., & Law, R. (2008). Progress in information technology and tourism management: 20 years on and 10 years after the internet—The state of etourism research. *Tourism Management, 29*(4), 609–623.

Burke, B. (2014). *Gamify: How gamification motivates people to do extraordinary things*. Brookline, MA: Bibliomotion.

Gebhardt, L., Krajzewicz, D., Oostendorp, R., Goletz, M., Greger, K., Klötzke, M., ... Heinrichs, D. (2016). Intermodal urban mobility: Users, uses, and use cases. *Transportation Research Procedia, 14*, 1183–1192.

Geneco. (2016). Case study: Bio-Bus. Retrieved December 21, 2017, from http://www.geneco.uk.com/Case_study_bio_bus/.

Gillan, A. (2014). Taiwan, home to the best street food markets in the world. Retrieved December 30, 2017, from https://www.theguardian.com/travel/2014/may/17/taiwan-taipei-street-food-markets.

Hanharan, J. (2010). Ecotourism and sustainability in the tourism sector. In L. Leonard & J. Barry (Eds.), *Global ecological politics* (Advances in Ecopolitics) (Vol. 5, pp. 171–229). Emerald Group Publishing Limited.

Hanley, S. (2015). Hilton Hotels partners with Tesla and GE to add charging stations. Retrieved December 26, 2017, from https://ecomento.com/2015/10/12/hilton-hotels-tesla-ge-electric-car-charging-stations/.

Harris, B. (2017). Tesla's electric truck 'needs the energy of 4,000 homes to recharge. Retrieved December 28, 2017, from https://www.weforum.org/agenda/2017/12/tesla-s-electric-truck-needs-the-energy-of-4-000-homes-to-recharge-say-researchers//.

Hernandez, D. (2016). Las Vegas casinos seek to power their bright lights with renewable energy. Retrieved December 14, 2017, from https://www.theguardian.com/environment/2016/mar/07/las-vegas-casinos-solar-power-nevada-energy.

Herrera, C. F, Herranz, J. B., & Arilla, J. M. P. (2012). Gastronomy's importance in the development of tourism destinations in the world. United Nations World Tourism Organisation. Global report on Food Tourism (pp. 6–9). Madrid: UNWTO.

Hirtenstein, A. (2017). London's iconic red buses to run on coffee in bid to cut emissions. Retrieved December 29, 2017, from http://www.independent.co.uk/news/business/news/london-red-buses-run-coffee-biofuel-cut-emissions-vehicles-air-pollution-biobean-shell-a8064516.html.

Kaplan, A. M., & Haenlein, M. (2010). Users of the world, unite! The challenges and opportunities of social media. *Business Horizons, 53*(1), 59–68.

Kastelein, B. (2004). Urban ecotourism: Impossible conundrum? Mexican cities must clean up their acts to keep tourists. *Business Mexico*, pp. 36–42.

Kazakova, K. (2015). Tourism gamification examples and what is there for you? Retrieved December 22, 2017, from https://www.ojoo.com/tourism-gamification-examples-and-what-is-there-for-you/.

Kramer, M. (2009). Soft mobility—Measures for a climate-friendly transport policy in Europe. *The Greens | EFA in the European Parliament.* Brussels, Belgium.

Merrick, J (2011). Glasgow and Edinburgh: The architectural rivalry. Retrieved December 13, 2017, from http://www.independent.co.uk/arts-entertainment/architecture/glasgow-and-edinburgh-the-architectural-rivalry-2315854.html.

Negruşa, A. D., Toader, V., Sofică, A., Tutunea, M. F., & Rus, R. V. (2015). Exploring gamification techniques and applications for sustainable tourism. *Sustainability, 7*, 11160–11189. https://doi.org/10.3390/su70811160

Nualkhair, C. (2015). Bangkok's best street food: A guide to dishes and districts. Retrieved December 30, 2017, from https://www.theguardian.com/travel/2015/nov/25/bangkok-best-street-food-guide-dishes-districts-thailand.

OECD. (2012). Climate change. Retrieved July 15, 2019, from https://www.oecd.org/env/cc/Work-on-Climate-Change-2013-14_web.pdf.

Özdemir, G., & Çelebi, D. (2015). Reflections of destinations on social media. In V. Katsoni (Ed.), *Cultural tourism in a digital era* (Springer Proceedings in Business and Economics). Cham: Springer.

Padilla, M (2017). A plug for electric car rentals. Retrieved December 27, 2017, from http://www.travelweekly.com/North-America-Travel/Insights/A-plug-for-electric-car-rentals.

Peltier, D. (2015). The future of food tourism goes beyond the restaurant experience. Retrieved December 10, 2017, from https://skift.com/2015/10/23/the-future-of-food-tourism-goes-beyond-the-restaurant-experience/.

Peltier, D. (2017). Local food trend keeps farms at center of tourism strategies. Retrieved December 10, 2017, from, https://skift.com/2017/04/25/local-food-trend-keeps-farms-at-center-of-tourism-strategies/.

Rapier, R. (2017). U.S. electric vehicle sales soared in 2016. Retrieved December 20, 2017, from https://www.forbes.com/sites/rrapier/2017/02/05/u-s-electric-vehicle-sales-soared-in-2016/#697d2a22217f

Roberts, S., & Tribe, J. (2008). Sustainability indicators for small tourism enterprises—An exploratory perspective. *Journal of Sustainable Tourism, 16*(5), 575–594.

Rocca, R. A. L. (2009). Soft mobility and urban transformation: Some European case studies. *TeMALab Journal of Mobility, Land Use and Environment, 3*, 85–80.

Rokka, J., & Moisander, J. (2009). Environmental dialogue in online communities: Negotiating ecological citizenship among travellers. *International Journal of Consumer Studies, 33*(2), 199–205.

Rowell, C. (2017). Gardens by the Bay, Singapore, breaks new ground in sustainable building. Retrieved December 17, 2017, from http://www.construction-global.com/major-projects/gardens-bay-singapore-breaks-new-ground-sustainable-building.

Saarinen, J. (2006). Traditions of sustainability in tourism studies. *Annals of Tourism Research, 33*(4), 1121–1140.

Santos, S. (2011). 2012 social media and tourism industry statistics. Retrieved December 22, 2012, from http://www.stikkymedia.com/blog/2012-social-media-and-tourism-industry-statistics.

Sarkar, S. K., Au, N., & Law, R. (2013). Analyzing ecotourists' satisfaction in socialization and knowledge sharing intentions via social media. In *Information and communication technologies in tourism* (pp. 313–326). Springer International Publishing.

Seraj, M., & Ayesugul, T. (2012). Social network citizenship. In M. M. Cruz-Canha, P. Goncalves, N. Lopez, E. M. Miranda, & G. D. Putnik (Eds.), *Handbook of research on business social networking: Organizational, managerial and technological dimensions* (pp. 339–357). IGI Global.

Sever, N. S., Sever, G. S., & Kuhzady, S. (2015). The evaluation of potentials of gamification in tourism marketing communication. *International Journal of Academic Research in Business and Social Sciences, 5*(10), 188–202.

Siemens AG. (2013). Annual report on sustainability. Retrieved July 15, 2019, from https://www.siemens.com/annual/13/en/download/pdf/Siemens_AR2013.pdf.

Tate, Z. (2017). Capitalism is losing support. It is time for a new deal. Retrieved December 30, 2017, from https://www.weforum.org/agenda/2017/11/capitalism-losing-support-we-need-a-new-deal/.

Transdanube. (2014). *Transnational soft mobility and tourism marketing strategy for transdanube regions*. Danube Competence Center.

UNWTO. (2004). Sustainable development of tourism. Retrieved December 26, 2017, from http://sdt.unwto.org/content/about-us-5.

UNWTO. (2012). Global report on city tourism—Cities 2012 Project. UNWTO, Madrid.

UNWTO. (2017). The tourism sector highlights the potential of urban tourism and the need to move toward more sustainable practices. Retrieved January 5, 2018, from http://media.unwto.org/press-release/2017-05-12/tourism-sector-highlights-potential-urban-tourism-and-need-move-toward-more.

WCED. (1987). *Our common future*. World Commission on Environment and Development. Oxford: Oxford University Press.

WEF. (2017). Technology and innovation for the future of production: Accelerating value creation. *World Economic Forum White Paper*.

Index[1]

[1] Note: Page numbers followed by 'n' refer to notes.

B. George, J. Paul (eds.), *Digital Transformation in Business and Society*, https://doi.org/10.1007/978-3-030-08277-2

Lightning Source UK Ltd.
Milton Keynes UK
UKHW022144081019
351244UK00003B/177/P

9 783030 082765